Design & Operation of Passenger Ships 2019

London, United Kingdom
1 May 2019

ISBN: 978-1-5108-9540-9

Printed from e-media with permission by:

Curran Associates, Inc.
57 Morehouse Lane
Red Hook, NY 12571

Some format issues inherent in the e-media version may also appear in this print version.

The Royal Institution of Naval Architects
8-9 Northumberland Street
London, WC2N 5DA
United Kingdom

Phone: 020 7235 4622
Fax: 020 7259 5912

publications@rina.org.uk

Additional copies of this publication are available from:

Curran Associates, Inc.
57 Morehouse Lane
Red Hook, NY 12571 USA
Phone: 845-758-0400
Fax: 845-758-2633
Email: curran@proceedings.com
Web: www.proceedings.com

TABLE OF CONTENTS

SHIP EVACUATION AND EMERGENCY RESPONSE TRENDS 1
F. Stefanidis, E. Boulougouris, D. Vassalos

ASSESSMENT OF THE MINIMUM SPEED REQUIREMENT IN SAFE RETURN TO PORT 8
R. Grin, R. De Meij, S. Rapuc

CONCEPT OF DESIGN AND IMPLEMENTATION OF PILGRIM SHIPS IN INDIA 16
Y. Kamath, K. Sivaprasad

DESIGN FEATURES OF RIVER-SEA CRUISE PASSENGER VESSELS FOR RUSSIAN INLAND WATERWAYS AND ADJACENT SEAS 20
A. Egorov, G. Egorov, I. Ilnitskiy

INTRODUCTION OF NEW SHAPED SHIP HULL "Y-HULL" 29
A. Wye

AN EXPERIMENTAL INVESTIGATION INTO REDUCTION OF SLOSHING EFFECTS IN SWIMMING POOLS ONBOARD CRUISE VESSELS 36
D. Nunes, D. Hudson

SPEED: A NEW SOFTWARE FOR SPEED & POWER SEA TRIAL ANALYSIS 48
P. Becchi

COMPUTATIONAL ANALYSIS TO CONTROL WELDING DEFORMATION IN THIN PLATE 55
M. Ali

Author Index

Ship Evacuation and Emergency Response Trends

F Stefanidis, Maritime Safety Research Centre, University of Strathclyde, UK
E Boulougouris, Maritime Safety Research Centre, University of Strathclyde, UK
D Vassalos, Maritime Safety Research Centre, University of Strathclyde, UK

SUMMARY

Accidents such as the Costa Concordia and more recently the Viking Sky incident cause a societal pressure for improving safety and emergency response in passenger ships. Finding realistic solutions for improvement requires first and foremost an understanding of the current regulatory landscape and the corresponding performance assessment standards. The first part of this paper is dedicated to the provision of a comprehensive outline of the regulatory framework that will ensure compliance of any new system and model developed. The second part is dedicated on the state-of-art projects and novel ideas on ship evacuation analyses, Life Saving Appliances (LSAs), Search and Rescue (SAR) with the purpose of unveiling areas for improvement. Finally, having identified the gaps in the aforementioned topics, suggestions are made on how future work can address the challenges of marine accident response.

1. INTRODUCTION

1.1 PASSENGER SHIP SIZE TRENDS

During the last decade, the tourist industry has shown a rapid growth in the demand of cruises that has led to an increase of the average cruise ship size by 30% [1-3]. This was driven mainly by the effort to make the most of the profit opportunities that come with the economies of scale [4] while trying to meet the market demand which continuously exceeded supply [5, 6].
This increase in passenger capacity is in line with a transformation on the cruise ship companies' business models that, in the same decade, have seen their ticket prices dropping due to the price elastic demand and competition [2]. Hence, companies focus on increasing their on-board revenue to make their reliance on ticket revenue smaller. This has changed the cruising experience over the years making the cruise ship itself the destination.
Financial drivers such as the ones above, together with port accessibility limitations have led to vessels with increased deck numbers and large entertainment areas spanning over many ship fire zones. This kind of design changes have brought safety related issues to the spotlight of technical challenges. Forming the "last resort" in case of an emergency onboard and its time-bound nature, ship evacuation and its corresponding procedures, modelling techniques and regulations should be constantly reviewed and updated.

1.2 PASSENGER SHIP SAFETY

Maritime Safety has been the cornerstone for the establishment of the International Maritime Organisation (IMO) and the introduction of its Safety of Life at Sea Convention (SOLAS). Both, passenger ships under operation and new builds, should guarantee the safety of the passengers and crew and that sometimes means going beyond the regulatory safety requirements. Unfortunately, accidents such as the fire on MS Norman Atlantic [7], the grounding of Costa Concordia and more recently the Viking Sky incident underpin gaps in our existing level of safety and emergency response. The societal pressure after incidents like these is forcing the regulatory bodies to act. IMO has introduced the Formal Safety Assessment [8] as a yardstick to quantify risks and gaps in the existing regulatory framework and select cost-effective risk control options (RCOs).
Safety is a life-cycle risk management endeavour, covering design, operation and emergencies [9]. Regretfully, safety had been perceived in the past as a constrain when it would be more beneficial to be treated as an objective [10]. This approach started gaining more and more ground at the end of 90s under the concept of Risk-Based Design (RBD). In the regulatory framework this was reflected by the introduction of the Goal Based Standards (GBS) and the Alternative Design and Arrangements (AD&A) [11]. Both are provisions to allow innovative designs to overcome bottlenecks emerging from the, frequently conflicting, design objectives of performance, functionality and safety. For the emergencies, we have seen the introduction of evacuation simulations and the introduction of the Safe Return to Port (SRtP, MSC.216(82)).
This kind of proactive risk management thinking should be overarching, from the design stage to the operation and emergency stage. In other words, there should be systems in place that can quantitatively assess the safety level of the vessel in real time and provide the decision makers with comprehensive situation awareness and a support for decision-making. The use of properly and real-time sensor-fused Decision Support Systems, able to provide feedback and guidance to all levels of actors during an emergency, can really revolutionize the evacuation process and create a step change in emergency response.
During the last decade, key enablers were the technological advancements, especially in the field of wireless communications and smart devices and the development of very accurate ship evacuation analysis models. Furthermore, the innovation achieved by a number of projects funded by the European Commission targeting tangible improvements to maritime safety have made brought us closer to the development of a DSS with unprecedented capabilities.

2. REGULATORY FRAMEWORK

2.1 EVACUATION

Currently passenger ship evacuation regulatory conformity requires compliance with the IMO MSC.1/ Circ. 1533 [12] which constitute the revised guidelines for evacuation analysis for both new and existing passenger ships. Before Circ.1533, the MSC.1/Circ.1238 [13] required evacuation analysis to be conducted only on the escape routes of ro-ro ships, whereas for the rest of passenger ships, the guidelines where applied voluntarily or for the purpose of demonstrating 'equal level of safety' for approval under the AD&A provisions. Accidents in cruise ships though, and experience gained via evacuation analyses brought to light the fact that voluntary compliance is inadequate and hence the Resolution MSC.404 (96) [14] was adopted that proposes amendments to SOLAS Regulation II-2/13 'Means of escape' [15]. These amendments will make the evacuation analysis compulsory not only to ro-ro ships but also to any other new passenger ships constructed on and after 1st of January 2020. This analysis is of paramount importance, especially for the state-of-the-art passenger ships where passenger capacity, practical challenges and large communal entertainment areas give rise to congestion points and critical areas, thus requiring a revision of the operational counter-measures for a timely evacuation.

The regulations for ship evacuation are based, at the moment, solely on the FSS code [16] and set the performance standard to 60 minutes for ro-ro vessels and passenger ships with less than three main vertical zones whereas for vessels with more than three vertical zones the limit is 80 minutes. In any case, the Embarkation and Launching time is considered to take up to 30 minutes. Those time limits should be checked with regard to some benchmark scenarios. In addition, the regulations provide a step-by-step procedure for a simplified evacuation analysis that models the problem as a hydraulic network, with the corridors and staircases being the pipes, the doors being treated as valves and the public spaces acting as tanks [12]. The assumptions of the simplified method are very limiting and, although it might provide an approximation on the expected evacuation time, it is unreliable and not so easy to use as the passenger demographics become more diverse, the accommodation more complex and the number of stairways and decks increases.

Fortunately, the regulations also provide sets of characteristics and parameters that an advanced evacuation method should take into account. An advanced model must calculate the individual time to evacuate in mixed passenger demographics in which the abilities of the passengers are distributed probabilistically [17], their movements are recorded and there is a universal algorithm that defines the rules of personal decision amongst the passengers. Finally, the same parameters that are used for the evacuation modelling in other fields, i.e. Geometrical, Population, Environmental and Procedural, should also be considered in advanced modelling. In the case of a vessel, that translates as the capability of the analysis to account for the different escape routes and their availability, initial passenger and crew distribution and their corresponding moving speed abilities. More specifically though, the condition of the vessel (static and dynamic) and the effect of crew assistance in the case of emergency.

ISM Code Sec. 7&8 and Resolution A.1072(28) cover the evacuation procedures, drills and exercises whereas the officers' training and responsibilities in crisis management and emergency cases are outlined in the STCW Code Sec. AII & AV [18, 19]. In addition, SOLAS Ch. III also includes information regarding the safety instructions to passengers and the characteristics of the Decision Support System that should be in place.

2.2 LSAs

The regulatory framework for Life Saving Appliances (LSAs) is being shaped initially by SOLAS Ch. III and the LSA Code which assess the design of the LSA against the requirements, identify potential deviations [15, 20]. Physical tests are then performed to ensure compliance of the non-deviating characteristics according to Resolution MSC.81 (70) [21]. As for the deviating characteristics of the LSAs, the proof of design equivalence falls under the regulations of SOLAS III-Reg.38 and the corresponding guidelines from AD&A provisions [11, 15], whereas the physical testing requirements of the deviating characteristics are prescribed by the Resolutions A.520(13) and A.689(17) [22, 23].

2.3 SAR

There also regulations in place for the procedures of alerting the Search and Rescue authorities [24] in case of an emergency as well as daily reports to the ship company informing them about the vessel's position [25]. The helicopter landing areas and the co-operation plans between SAR and Passenger ships are also subjects of IMO legislation [26-29].

Table 1: Regulatory reference by subject

Subject	Regulatory Reference	
Evacuation Modelling/ Procedures	MSC.1/Circ.1533	[12]
	MSC. 404(96)	[14]
	SOLAS 2009	[15]
	FSS Code	[16]
	ISM Code	[18]
	A.1072(28)	[19]
Life Saving Appliances	*SOLAS 2009*	[15]
	LSA Code	[20]
	MSC.81(70)	[21]
	A 689(17)	[22]
	A.520(13)	[23]

Search and Rescue	MSC/Circ.892 MSC/Circ.1043 MSC/Circ.895 MSC.1/Circ.1079/Rev.1 COMSAR/Circ.31 IAMSAR	[24] [25] [26] [27] [28] [29]

3. STATE OF THE ART LSAs

3.1 LIFEBOATS

It is worth mentioning, that although LSA Code 4.4.2.1 explicitly states that there cannot be a lifeboat of capacity greater than 150 people, the AD&A provisions have allowed the creation of 'mega- lifeboats' that have received approval after demonstrating equivalent level of safety. European commission projects such as SAFEDOR and SAFECRAFTS have played a crucial role in the realization and commercialization of such lifeboats [30, 31].

Most noticeable being the 370-person CRV55 lifeboats from Umoe Schat-Harding that are installed in Oasis of the Seas [32]. These catamaran hull shaped 'mega-lifeboats' are designed to serve the needs of the new 'mega-liners' where the exponential increase in passenger capacity and the available deck area for LSAs makes the use of conventional lifeboats impractical or even impossible. The launching system for these boats, the LS45 davit, is also considered novel since it is designed specifically for cruise vessels and allows for the lifeboats to be lowered to water directly without any outswing since the boats are kept outside the hull. Finally the catamaran hull choice has allowed for improved seakeeping performance. Another example is the SEL-T 15.5 by Fassmer which was the first IMO certified lifeboat offering a persons capacity up to 267 while also being certified for usage as a tender for 233 passengers.

Innovation in lifeboats is now focused on creating different release mechanisms and davits with as few moving parts as possible so that less maintenance to be required. Another tendency driven by the cruise industry is the ability of the lifeboats to be used also as tenders, thus leading to designs with ample space, window view while maintaining the robustness and reliability expected from a lifeboats [33]. Also, efforts are made towards reducing the complexity in both the launching and operational phase of these boats so that no much training to be necessary. Further development, though, is still needed in making these boats available for launching and use even after extreme heeling angles or releases from considerable heights. Ensuring the maximum availability of all the boats and LSAs in general requires also a careful selection on their location at the vessel. Damage stability and fire analysis should indicate the most vulnerable areas on-board and the location of the lifeboats should be determined by taking these factors under consideration.

Lifeboats occupy approximately 0.05% of a passenger ship´s volume but they do represent a 1%-2% of the total ship building cost and therefore ensuring cost effective and safe alternatives to the conventional boats is essential [34]. What also adds a certain level of challenge to the design of the lifeboats is the pressure from ship owners of increasing the number of ocean view cabins, which generate more profit, while also not taking over much deck area and do not obstruct the view.

3.2 LIFE-RAFTS AND MES

Marine evacuation systems are used to ensure the evacuation of the maximum number of passengers in the minimum amount of time and securely place them to the already deployed life rafts. The most common types of Maritime Evacuation Systems (MES) for fast and easy evacuation in large passenger ships are the single or double chute systems such as the VEDC by VIKING and the Marin Ark 2 by Survitec (RFD) [35, 36]. The manufacturers claim that these systems are capable of evacuating more than 850 passengers in less than 30 minutes and inflated within approximately 90 seconds after deployment. Those systems are designed to securely transfer persons from deck heights up to 16.8 meters, via helical slide paths in fully enclosed chutes, into the life rafts. Designed to require minimal human element involvement but, more importantly, have a flexible design that allows installation in wide range of vessel configurations.

Smaller passenger vessels with low embarkation decks can make use of more compact lightweight MES such as Mini Chute Systems, Slide systems or direct boarding life rafts a few examples of which are VEMC, VEC Plus from VIKING and their equivalent Bruce Evacuation System Chute and Super Slide from Survitec.

The next generation of MES, similarly with the next generation of lifeboats should ensure the same or higher performance standards in trim and listing conditions and decrease the deck footprint. In addition, as arctic cruises become more and more popular, the heat insulation of these systems should also be reviewed.

3.3 LIFEJACKETS AND SMART DEVICES

Effective decision making for emergency depends upon the level of information that the ship officers have about the actual real time state of damage and passenger distribution within the vessel. Lack of sufficient and reliable information on these issues put on more strain to the decision makers [37] and directly affect the efficiency of the evacuation and SAR.

The research outputs of European Commission funded project LYNCEUS managed to demonstrate the benefits of using smart devices, such as wireless bracelets and lifejacket-embedded sensors, in ship evacuation. More specifically the project investigated how ultra-low power wireless area network technologies can be utilised for people localisation during emergencies [38]. This kind of localisation capabilities open new horizons for real-time emergency response management. These smart devices, lifejackets, the associated software and Decision Support

System are being brought to market via the second phase of the project 'Lynceus2Market' [39]. Lifejacket innovation can also come from technologies developed for sports [40] or military applications that can introduce more compact designs that also include survival kits.
Nevertheless, there are still many technical challenges to be addressed related to passenger health status monitoring, communication system protocols due to the large number of wireless sensors and bandwidth availability, all of which can play a key role in the evacuation. Especially for the matter of indoor localization, feedback from projects such as FP7 eVACUATE suggest the exploration of Ultra-Wide Band (UWB) technologies.

4. SHIP EVACUATION MODELLING

Ship evacuation modelling is significantly more complex than building evacuation because of, amongst other reasons, unique challenges arising from the fact that passengers cannot directly head towards the nearest escape exit but instead should be equipped with lifejackets, available on their cabins and/or elsewhere, and prepare themselves for embarkation in LSAs that are going to transfer them from one high risk area to another high risk environment, the sea. The complexity of modelling gets even harder when considering that most of the passenger ships have rambling layouts, identical corridors and confusing layout diagrams that make the passenger orientation problem worse in an already stressful environment [41]. Therefore, there are procedural, human behaviour and environmental factors that constitute conventional evacuation software, developed for other industries, unsuitable for marine emergency evacuation.
As of now, the most prominent ship evacuation modelling software are Evi [42], maritimeExodus [43, 44], IMEX [45], AENAS [46, 47], VELOS [48] each one developed to evaluate the time to evacuate using different modelling methods, taking different factors under consideration and different assumptions.
Evi is a multi-agent based model capable of representing alignment and cohesion behaviours by adjusting the agent speed depending on the density of agents in a space [49, 50]. The multi-agent approach of the software places it amongst the most appropriate models for multi-level planning structure. In the case of fire accident, the maritimeEXODUS software can simulate the fire and smoke propagation together with the passenger evacuation using velocity-based fine network mode. ANEAS and IMEX were created to solve problems arising from the consideration of ship motions in the evacuation modelling via velocity reduction coefficients for different deck inclinations [51, 52]. VELOS brought the Virtual Reality into the evacuation simulation with the purpose of creating a platform that would allow for design feedback at the early stages via the immersion of multiple users into hectic operational conditions.
Nevertheless, there is still room for improvement on not only the human behaviour under panic and the corresponding changes in the walking speed, but also the effect of disabled people on the evacuation flow. From the standpoint of environmental considerations, the effect of trim and heeling angles in the availability of the LSAs and the corresponding effects on passenger flow are still to be modelled. More importantly, though, there is not a dynamic enough model, yet, capable of assessing the total time to evacuate based on real time passenger localisation data. The real time coupling of passenger tracking and evacuation time calculation is a crucial gap, without which the ability of creating a Decision Support System that will allow the optimal co-ordination of the evacuation process by assessing the most favourable escape route for each individual.

5. EVACUATION ENHANCEMENT

Starting with the LSAs, a new generation of cost effective personal survival equipment must be developed together with innovative concepts in ship design layout that can accommodate novel lifeboat designs and increase the evacuation efficiency by considering the spatial constraints in place. Effectively this means more compact and ergonomic life-jackets, fitted with the necessary sensors for localisation and pairing abilities with health monitoring and information receiving devices such as smart hand bands and earplugs. Augmented Reality (AR) technology could also be employed for the secure guidance of the passenger throughout the evacuation stage. Alternative ship layouts should be examined based on their efficiency in reduction of the time for mustering and abandonment.
In terms of ship evacuation modelling, the new challenges and targets will be the incorporation of real time data, from the various sensors monitoring both the type and propagation of damage and human physiological factors, in an evacuation analysis model that can evaluate different route alternatives for individuals while taking the human behaviour under consideration [53]. Evacuation models can also be improved with regards to their accuracy in embodying the differences in walking speed as a function of the individuals' specific psychological and physical characteristics, that means to include family and panic behaviour as well as the effect of disabled people in the evacuation flow [54]. Moreover, the recent acts of terrorism have added the security threat in the list of subjects that should be investigated with respect to evacuation [55, 56]. Another area of improvement is the calculation of the actual time for lifeboat embarkation. Most of the models calculate explicitly only the time to evacuate up until the assembly station and are not able to do the same for the lifeboat boarding stage due mainly to the lack of operational or experimental data that would allow verification.
Besides the aforementioned suggestions for evacuation enhancement, recent studies on the combined effect of trim and heeling angles on the walking speed [57] allow for refinements on the accuracy of existing models and pave the way for a truly more dynamic evacuation

analyses. The continuous, non-static, effect of ship motions in walking speeds has also to be linked with real time flooding simulations so that the time-to-evacuate can be associated with the time-to-capsize, which is an important connection to be made in the next generation of evacuation modelling software.
The biggest challenge though, will be the integration of all the systems related to the evacuation and emergency response under one Decision Support System that will broadly cover all the emergency cases and co-ordinate the evacuation process more efficiently. Data from flooding and fire sensors should quantify the risk of untenable propagation of damage in real time while also giving to the decision makers on the actual available time for evacuation based on passenger localisation data. In the meantime the risk of evacuation itself according to the available search and rescue options should be evaluated to lead the officers to a well informed decision and reduce the human errors. This task is particularly challenging in view of the technical difficulties that have to be overcome and the number of different stakeholders that have to work together to produce cost effective solutions that can be integrated with the software and hardware products of other companies or research groups.

6. CONCLUSIONS

Since ship evacuation is a dynamic multi-variable problem with parameters that are constantly changing and evolve as a function of time, environmental conditions, human behaviour, and type of emergency, it is essential to have a dynamic evacuation analysis model. Dynamic in the sense that can effectively calculate the available time to evacuate (ASET) and the required time to evacuate (RSET) based on the state and location of fire, flooding or security threat, for instructing accordingly the passengers on how to proceed. Improving ship evacuation is a problem that has to be approached simultaneously from all angles. Novelty in the LSAs, real time evacuation analysis and the creation of a Decision Support System that will quantify in real time the risk of evacuation and weight it against the risk of ship loss and the number of potential fatalities are being the challenges of the upcoming decade.
The feasibility of the aforementioned system is a matter that requires the collaboration and co-ordination of academic research groups with passenger ship companies, LSA manufacturers and software and tele-communication engineering teams. This kind of initiatives become possible after the financial support and encouragement by regulatory and governmental bodies for research and innovation actions. For instance, European Commission funded projects under the call for marine accident response initiatives (MG-2-2-2018) are going to bring together academia and industry to develop and test systems and devices that will improve evacuation. An example of such a project is SafePASS that aims to design the next generation of life saving appliances and systems for a safer and more efficient evacuation of large passenger ships in extreme emergency and environmental conditions.

7. AKNOWLEDGMENETS

The authors greatly acknowledge the sponsorships of DNVGL and Royal Caribbean Cruises Ltd. to the Maritime Safety Research Centre at the University of Strathclyde.
The opinions expressed herein are those of the authors and do not reflect the views of DNVGL and RCCL.

8. REFERENCES

1. CLIA, 'Cruise Industry Outlook', *Cruise Lines International Association*, 2016.
2. Dowling, R et al., 'Cruise Ship Tourism', *CABI*, 2017.
3. Tsitsilonis, K.M., et al., 'Concept Design Considerations for the next generation of Mega-Ships', in *13th International Marine Design Conference (IMDC 2018)*, 2018
4. Chang, Y.T, Lee S.H. et al., 'Efficiency analysis of major cruise lines', *Tourism Management*, 2017.
5. FCCA, 'Cruise Industry Overview 2016', *Florida- Caribbean Association*, 2016.
6. Institute of Shipping Economics and Logistics, 'Shipping Statistics and Market Review', *Volume 61-No. 5/6*, 2017.
7. Directorate General for Rail and Marine Investigations, 'Fire on-board the ro-pax NORMAN ATLANTIC', *Ministry of Infrastructure and Transport*, 2014.
8. International Maritime Organization (IMO), 'MSC-MEPC.2/Circ.12: Revised Guidelines for Formal Safety Assessment (FSA) for use in the IMO rule-making process', *IMO*, 2013.
9. Vassalos, D., 'Design for safety, risk-based design, life-cycle risk management', *RINA Annual General Meeting*, 2014.
10. Papanikolaou, A., et al, 'Risk-Based Ship Design: Methods, Tools and Applications', *Springer*, 2009.
11. International Maritime Organization (IMO), 'Guidelines on Alternative Design and Arrangements for SOLAS Chapters II-1 and III', *IMO*, 2006.
12. International Maritime Organization (IMO), 'MSC.1/Circ.1533: Revised guidelines on evacuation analysis for new and existing passenger ships', *IMO*, 2016.
13. International Maritime Organization (IMO), 'MSC.1/Circ.1238: Guidelines for Evacuation Analysis for New and Existing Passenger ships', *IMO*, 2007.
14. International Maritime Organization (IMO), 'Resolution MSC. 404(96): Amendments to the

international convention for the safety at sea', *IMO*, 1974.
15. International Maritime Organization (IMO), 'SOLAS- International Convention for the Safety of Life at Sea', *IMO*, 2009.
16. International Maritime Organization (IMO), 'FSS Code: International Code for Fire Safety Systems', *IMO*, 2015.
17. Galea, E.R., et al., 'Recommendations on the Nature of the Passenger Response Time Distribution to be used in the MSC.1033 Assembly Time Analysis Based on Data Derived from Sea Trials', *International Journal of Maritime Engineering*, *IMO*, 2007.
18. International Maritime Organization (IMO), 'International Safety Management (ISM) Code', *IMO*, 2015.
19. International Maritime Organization (IMO), 'Resolution A.1072(28): Revised guidelines for a structure of an integrated system of contingency planning for shipboard emergencies', *IMO*, 2013.
20. International Maritime Organization (IMO), 'LSA Code-MSC.48(66): International life-saving appliance code', *IMO*, 1998
21. International Maritime Organization (IMO), 'MSC.81(70): Revised Recommendation on Testing of Life-saving Appliances', *IMO*, 1998.
22. International Maritime Organization (IMO), IMO, 'Resolution A 689(17): Testing of life-saving appliances', *IMO*, 1996.
23. International Maritime Organization (IMO), 'Resolution A.520(13): Code of Practice for the evaluation, testing and acceptance of prototype novel Life-saving appliances and arrangements', *IMO*, 1983.
24. International Maritime Organization (IMO), 'MSC/Circ.892: Alerting of Search and Rescue Authorities', *IMO*, 1998.
25. International Maritime Organization (IMO), 'MSC/Circ.1043: Guidance on ship's daily reporting of their positions to their companies', *IMO*, 2002.
26. International Maritime Organization (IMO), 'MSC/Circ.895: Recommendation on helicopter landing areas on Ro-Ro passenger ships', *IMO*, 1999.
27. International Maritime Organization (IMO), 'MSC.1/Circ.1079/Rev.1: Guidelines for preparing plans for cooperation between search and rescue services and passenger ships', *IMO*, 2017.
28. International Maritime Organization (IMO), 'COMSAR/Circ.31: Guidance for mass rescue operations', *IMO*, 2003.
29. International Maritime Organization (IMO), 'International Aeronautical and Maritime Search and Rescue (IAMSAR) Manual', *IMO*, 2016.
30. SAFEDOR, 'Design, Operation and Regulation for Safety', *European Commission- CORDIS*, 2009.
31. SAFECRAFTS. 'Safe abandoning of ships, Improvement of current Life Saving Appliances Systems', *European Commission- CORDIS*, 2012.
32. RINA, 'New 37-person mega lifeboat and davit system developed', *The Royal Institute of Naval Architects*, 2009.
33. Fassmer. 'SEL-T 15.5: Partially Enclosed Lifeboat/Tenderboat', *Fassmer GmbH & Co. KG*, 2018.
34. Papanikolaou, A., et al., 'Final Conference Proceedings', *SAFEDOR IP-516278: an Integrated project supported by the European Union*, 2009.
35. VIKING, 'VEDC: VIKING Evacuation Dual Chute', *VIKING Life- Saving Equipment A/S*, 2018.
36. Survitec Group, 'Marine Evacuation Systems: Providing Critical Safety Solutions', *Survitec Group Ltd*, 2017.
37. Kristiansen, S., 'Maritime Transportation-Safety Management and Risk Analysis', *Butterworth- Heinemann*, 2005.
38. Kounoudes, A., et al., 'Lynceus: People localisation for safe ship evacuation during emergency', *Annual Meeting of the Hellenic Institute for Maritime Technology*, 2012.
39. LYNCEUS2MARKET, 'An innovative people localisation system for safe evacuation of large passenger ships', *LYNCEUS2MARKET*, 2017.
40. ProVitaTec Ltd. 'Sens07vest: Floating system for free divers and spear fishers', *ProVitaTec Ltd*, 2018.
41. Ahola, M. and R. Mugge, 'Safety in passenger ships: The influence of environmental design characteristics on people's perception of safety', *Applied Ergonomics*, 2016.
42. Azzi, C., et al. 'Evacuation Simulation of Shipboard Fire Scenarios', *Fire and Evacuation Modeling Technical Conference*, 2011.
43. Glen, D.R., E.R. Galea, and e. al, 'Ship Evacuation Simulation: Challenges and Solutions', *Transactions-Society of Naval Architects and Marine Engineers*, 2001.
44. Gwynne, S., et al., 'Analysing the evacuation procedures employed on a Thames passenger boat using the maritimeEXODUS evacuation model', *Fire Technology*, 2003.
45. Park, J., et al., 'Development of Evacuation Model for Human Safety in Maritime Casualty', *Ocean Engineering*, 2004.
46. Meyer-Konig, T., H. Klupfer, and M. Schreckenberg, 'Assessment and Analysis of Evacuation Processes on Passenger Ships by Microscopic Simulation', *Pedestrian and Evacuation Dynamics*, 2002.

47. Meyer-Konig, T., P. Valanto, and D. Povel, 'Implementing Ship Motion in ANEAS-Model Development and First Results', *Pedestrian and Evacuation Dynamics*, 2007.
48. Ginnis, A.I., et al., 'VELOS: a VR platform for ship-evacuation analysis', *Computer Aided Design*, 2010.
49. Vassalos, D., et al., 'Advanced Evacuation Analysis-Testing the ground on ships', *2nd International Conference on pedestrian and evacuation dynamics*, 2003.
50. Vassalos, D., et al., 'A Mesoscopic Model for Passenger Evacuation in a virtual ship-sea environment and performance-based evaluation', *Pedestrian and Evacuation Dynamics*, 2002.
51. Sarvari, P.A., et al., 'Studies on emergency evacuation management for maritime transportation', *Maritime Policy & Management*, 2017.
52. Balakhontceva, M., et al., 'Multi-agent Simulation of Passenger Evacuation from a Damaged Ship under Storm Conditions', *The International Conference on Computational Science (ICSS)*, 2016.
53. Boulougouris, E. and A. Papanikolaou, 'Modeling and Simulation of the Evacuation process of Passenger Ships', *Proceedings of 10th International Congress of the International Maritime Association of the Midwinter (IMAM)*, 2002.
54. Miyazaki, K., et al., 'Evacuation simulation for disabled people in passenger ship', *National Maritime Research Institute*, 2004.
55. Gypa, I., E. Boulougouris, and D. Vassalos, 'Quantification of the maritime security problem onboard passenger ships', in *Transport Research Arena (TRA)*, 2018.
56. Gypa, I., E. Boulougouris, and D. Vassalos, 'Decision Support tool for modelling security scenarios onboard passenger ships', *Maritime Transportation and Harvesting of Sea Resources*, 2017.
57. Kim, H., M.I. Roh, and S. Han, 'Passenger evacuation simulation considering the heeling angle change during sinking', *International Journal of Naval Architecture and Ocean Engineering*, 2018.

9. AUTHORS BIOGRAPHY

Fotios Stefanidis is a PhD student in the department of Naval Architecture, Ocean and Marine Engineering in the University of Strathclyde. He graduated from the same department with an MEng degree in 2018.

Evangelos Boulougouris is Director of the MSRC. He is RCCL Reader of Safety of Marine Operations at the University of Strathclyde. His previous research and professional experience includes design for safety, multi-objective ship design optimisation, stability criteria, alternative fuels for ships, ship evacuation and the impact to safety and security. He is Fellow of RINA, member of the ITTC Technical committee on stability in waves and member of the STAB R&D committee for stability.

Dracos Vassalos is Professor of Maritime Safety at the University of Strathclyde and founder of the MSRC. He has received several awards including the Sustainable Achievement Award by the Royal Academy of Engineering for his contribution in Maritime Safety, the Froude Metal from the Royal Institute of Naval Architects and the David Taylor Medal from the Society of Naval Architects and Marine Engineers.

ASSESSMENT OF THE MINIMUM SPEED REQUIREMENT IN SAFE RETURN TO PORT

R Grin, R de Meij and **S Rapuc,** Maritime Research Institute the Netherlands (MARIN), The Netherlands

SUMMARY

The present paper discusses the aspects of how to assess the compliance of cruise ships and RoPax ferries with the requirement for Safe Return to Port as detailed in SOLAS Ch. II-2/21.4.1 – Propulsion. Part of this requirement is the necessity to be able to maintain a minimum sustained speed of 6 knots in Bft 8 weather conditions, for the worst single compartment damage case. This worst damage case is typically failure of one shaft line or engine room. The present paper discusses these methodologies applied to a 120m expedition cruise vessel. In addition, it presents the sensitivity of this requirement to exact environmental conditions, to the choice of blocked or water-milling propeller and to the ship design speed in calm water. In particular, it is demonstrated that for relatively small and slow passenger ships, the SOLAS requirement may necessitate to increase the intended onboard power.

NOMENCLATURE

Ax	Frontal wind-exposed area [m²]
Bft	Beaufort number [-]
CA	Correlation allowance [-]
Cx	Wind coefficient in x-direction [-]
FATIMA	MARIN's linear seakeeping potential-flow code
J	Advance coefficient [-]
KQ	Torque coefficient [-]
KT	Thrust coefficient [-]
N_{prop}	Propeller rotation rate [RPM]
Q_{prop}	Propeller torque [kNm]
ρ	Density of water [tonne/m^3]
ReFRESCO	MARIN's open-source CFD code
R_{wind}	Wind added resistance [kN]
SRtP	Safe Return to Port
V_{rel}	Relative wind velocity [m/s]
V_s	Ship speed [kn]
ρ_{air}	Density of air [tonne/m^3]
U_{wind}	Wind speed at 10 m altitude [m/s]
QTF	Quadratic transfer function
RAW	Wave added resistance [kN]
S_{psd}	Wave spectrum [m²s]
ω	Wave frequency [rad/s]
T_p	Peak wave period [s]
H_s	Significant wave height [m]

1. INTRODUCTION

Cruise ships and RoPax ferries, which are typically twin-screw ships, have to comply with the requirement for Safe Return to Port as detailed in SOLAS Ch. II-2/21.4.1 – Propulsion. Part of this requirement is the necessity to be able to maintain a minimum sustained speed of 6 kn in Bft 8 weather conditions for the worst single compartment damage case. This worst damage case is typically failure of one shaft line or engine room.
Over the years, at MARIN we have gathered experience on how to address this problem by means of calculation work, model tests and a combination of both. The present paper discusses these methodologies and some of the pitfalls to avoid, and presents the case study of a 120m expedition-type cruise vessel. Based on this case study, the paper also presents the sensitivity of this requirement to exact environmental conditions, to the choice of blocked or water-milling propeller, and to the ship design speed in calm water. In particular, it is demonstrated that for relatively small and slow passenger ships, the SOLAS requirement may necessitate to increase the intended onboard power. Finally, a case of bow quartering seas is presented.

2. METHODOLOGY

Passenger vessels have to comply with the requirement for Safe Return to Port (SRtP) as detailed in SOLAS Ch. II-2/21.4.1 – Propulsion. In particular, they must prove that they can maintain a minimum speed in Bft 8 wind and wave conditions with one propeller water milling and/or blocked. The present section discusses MARIN's methodology to assess a passenger ship's SRtP performance. The basis of this approach is that the calm water powering performance, the wind added resistance and the wave added resistance can be computed or measured separately and then combined to estimate the ship's sustained speed.

2.1 PROPULSION WITH BLOCKED/WATER-MILLING PROPELLER

The calm water performance with respect to the SRtP requirement can be obtained numerically, using e.g. empirical methods or more advanced Computational Fluid Dynamics (CFD) approaches. When following a numerical approach one should account for the additional drag of the water-milling or blocked propeller. It is also possible to measure the performance during a number of propulsion tests, typically for a speed range of 4 to 10 kn, which are carried out in calm water on a large ship model, as described below.

Firstly, a short propulsion test is carried out at a low speed range with both propellers working. Next, two series of propulsion tests are carried out with one propeller active. In the first series the other propeller is blocked (N_{prop}=0 RPM). This is the worst-case scenario due to the additional drag of the blocked propeller. In the second series, the propulsion test is carried out with one propeller water milling. This is a best-case scenario, as the shaft torque is assumed to be zero (no shaft bearing losses or engine torque; Q_{prop}=0 kNm). If the shaft bearing friction losses are known, then these values may be applied in this test. Lastly, a pulling test is carried out for the ship operating at a speed of 6 kn. During a pull test, the speed is fixed and the propeller rotation rates are increased until the maximum torque or power is reached. By doing this, the maximum thrust of the propeller is obtained at the correct thrust deduction fraction.

The same program may be applied during the assessment for a ship fitted with a twin or triple Azipod® propulsor arrangement. In this way, the most favourable and most unfavourable situations are tested for the ship with one engine room flooded or on fire.

In a full-scale operation, when the ship is driven by only one active propeller, for instance on starboard, this will result in the vessel starting to turn to port side. This turning moment is further increased by the additional drag of the water-milling portside propeller (assuming a blocked propeller makes this moment even larger). This turning moment has to be compensated by applying a steering angle to starboard. In the end, this situation will lead to the ship sailing at a certain drift angle over starboard with a non-zero rudder steering angle to starboard. While this drift angle and rudder angle are typically limited, they will cause some additional drag. However, such an equilibrium situation is very difficult and very expensive to realise in the basin, needing trial and error for a range of drift and rudder angles. For that reason and due to the fact that CFD calculations have shown that for most twin-screw ships the drift angle will be in the order of maximally 1 degree, these tests are usually done with the ship at 0 drift angle and using no rudder steering angle. This implies that the results of this test are somewhat more optimistic than what may be expected for the full-scale vessel in this condition.

In an earlier phase of the project, the open water characteristics of the propeller models will have been determined in propeller open water tests, using a so-called "quasi-steady" approach. This means that during each measurement run, both the forward speed and the propeller rotation rate are varied to cover the entire range of advance coefficients from $J = 0$ to $K_T = 0$. The propeller rotation rate has to be relatively low at low J-values to avoid overloading of the thrust and torque sensors. At higher J-values, the propeller rotation rate has to be as high as possible to minimise scale effects. The propulsor open water characteristics will be used in the analysis of the propulsion tests and for the estimation of the required power.

The results of the pull and propulsion tests are extrapolated to full-scale values according to the MARIN form factor method. In the MARIN 3-D extrapolation, use is made of the Grigson formula for model-ship correlation incorporating a form factor and a model-ship correlation allowance CA. The form factor 1+k represents the ratio of the viscous to the frictional resistance, and is based on resistance measurements at low speeds; it is verified with statistical data and with a CFD calculation. In the model-ship correlation allowance CA, the effects of the still-air drag of the wind exposed ship and superstructure size are included as well.

The values extrapolated according to this procedure refer to the so-called "tank" condition. To arrive at the shaft power prediction for ship trial conditions, scale effect corrections are applied to the advance velocity (wake) and to the propeller blade friction.

2.2 WIND ADDED RESISTANCE

The added resistance due to head wind is taken into account using the following formulation:

$$R_{wind} = \tfrac{1}{2} \cdot \rho_{air} \cdot C_x \cdot A_x \cdot \left(V_{rel}^{\,2} - V_s^2\right)$$

Where ρ_{air} is the density of air, C_x and A_x the frontal wind coefficient and frontal wind area, V_{rel} the relative wind speed (for head wind the summation of wind speed and vessel speed) and V_s the ship speed. Within the SRtP requirements, wind and waves are assumed to have the same direction. Therefore, only the component of the wind force in x-direction is taken into account. As the still-air drag is accounted for in calm water condition, this contribution is subtracted here. The frontal wind coefficient C_x is obtained from either wind tunnel tests or reference ships.

The wind speed for Bft 8 ranges from 17.2 to 20.7 m/s. For SRtP, typically the average is selected i.e. 19 m/s.

2.3 WAVE ADDED RESISTANCE

The ship wave added resistance can be either computed using advance seakeeping tools or measured during seakeeping model tests.

If it is computed, it is advised to use a forward speed Rankine source method such as FATIMA [1] that has been proven to give a fairly accurate estimate of the wave added resistance for a wide range of wave frequencies, wave heading and ship speeds. The resulting wave added resistance of such a tool is typically presented as Quadratic Transfer Function (QTF) as a function of the wave frequencies. This implies that the wave added resistance increase quadratically with the wave height. From a given QTF, it is possible to compute the mean wave added resistance in irregular waves using the wave spectrum (S_{psd}) and the following formula:

$$RAW = 2\int_0^{\infty} QTF(\omega) \cdot S_{psd}(\omega) \cdot d\omega$$

If the wave added resistance proves to be non-linear and/or when the Rankine source results are uncertain, it may be of interest to cross-validate the linear results for a few frequencies (typically around the peak and tail of the QTF) using CFD simulation tools such as ReFRESCO.

Alternatively, the wave added thrust in a given sea state can be obtained directly by means of seakeeping model tests. The typical approach is to perform free sailing tests with both propellers active in calm water and in the targeted sea state. By subtracting the obtained mean thrusts in both conditions, the wave added thrust is obtained directly. The main advantage of performing such model tests is that they include all physics and related non-linearities such as breaking waves, speed variations and propeller ventilation.

For the wave spectrum, typically standard spectra are applied. Common choices are JONSWAP for young growing seas (say Bft 8 for 2 hours on the North Sea) or Pierson-Moskowitz for fully developed seas (say Bft 8 for 2 days on the Atlantic Ocean).

Empirical information as used by Janssen et al. [2], relates the peak period T_p in [s] to the significant wave height H_s in [m] and wind speed U_{wind} at 10 m height in [m/s]. Their work can be expressed as:

$$T_p = \frac{9.25 \cdot H_s^{0.688}}{U_{wind}^{0.376}}$$

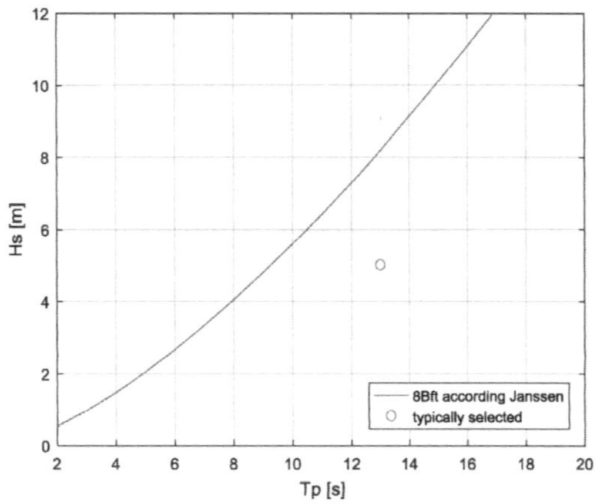

Figure 1: Relation between wave height, wave period and wind speed.

In recent SRtP projects, the most commonly adopted wave height was 5 m, with most of the higher and lower wave conditions within +/- 0.5 m. For the peak wave period the range is considerably larger; recently often a period of 13 s is chosen as suggested by some class societies, but this is not yet formalised. From Figure 1, however a Bft 8 condition tends to be typically in shorter waves of around 9 to 10 s.

2.4 COMPUTATION OF THE SUSTAINED SPEED

From the obtained calm water performance, wave added and wind added resistance it is possible to estimate the required thrust necessary to make a given speed considering one propeller water milling and/or blocked.

First, the computed wind and wave added resistance needs to be converted into thrust using a thrust deduction factor. This coefficient is typically obtained for the calm water pull and propulsion tests. When not available, 0.05-0.10 is a reasonable approximation for low Froude numbers. Note that if free sailing seakeeping model tests are done, the wave added thrust is directly measured.

Secondly, the maximum available thrust of the remaining propeller is obtained from the pulling test or can be calculated from the open water diagram. The maximum thrust is dependent on the available torque or power the main engine can deliver. Typically, a constant torque is assumed for the diesel direct propulsion and a constant power for a diesel electric propulsion plant. The latter gives a higher maximum thrust in overload condition.

Finally, by comparing this required thrust to the thrust available through the active propeller one can compute the ship's sustained speed in the given environmental conditions.

4. CASE STUDY

4.1 CRUISE SHIP DESIGN

A typical expedition-type cruise vessel design has been taken as reference ship (see Table 1). The twin-shaft vessel is equipped with two rudders, one stern thruster and two bow thrusters. Tables 2 and 3 indicate the main propulsion and propeller data of the ship.

Table 1. Main dimensions.

Quantity	Value	Unit
Length perpendiculars	120.0	[m]
Breadth	17.0	[m]
Draught	4.0	[m]
Displacement	5,800	[tonnes]
Metacentric height	1.7	[m]
Design speed	19	[kn]

Table 2. Propeller particulars.

Parameter	Value	Unit
Diameter	3.00	[m]
Pitch ratio at 0.7R	1.12	[-]
Boss-diameter ratio	0.211	[-]
Expanded blade area ratio	0.687	[-]
Number of blades	5	[-]

Table 3. Propulsion data.

Parameter	Description	Unit
Engine type	diesel with gearbox	[-]
Number and type of propulsors	2 fixed pitch propellers	[-]
Brake power	2 x 3750	[kW]
Propeller rotation rate	202	[RPM]
Shaft line & gearbox losses	1.0	[%]

4.2 ENVIRONMENTAL CONDITIONS

To get an impression of the wind and wave conditions encountered by cruise vessel, one can perform voyage simulations coupled to hindcast data. In particular, these simulations allow to obtained coupled wind and wave information along the cruise route.

For the present case study an expedition-type cruise vessel has been chosen sailing cruises in Antarctica and the Pacific islands in the summer of the southern hemisphere. When heading to the northern hemisphere, one or two longer "world" cruises are done. When arrived on the northern hemisphere, she will be mainly sailing in the Arctic area until the summer ends. Because of this sailing profile, the weather conditions are relatively harsh. The wind and wave conditions are obtained from hindcast data along the sailing route for a period of 10 years, provided by the European Centre for Medium-Range Weather Forecasts (ECMWF).

In the SRtP minimum speed requirement the wind speed is set at Bft 8. For this particular cruise scenario, the chance of having Bft 8 is slightly less than 1%, see Figure 2. Bft 9 or 10 conditions do occur on this route in summer but quite infrequently.

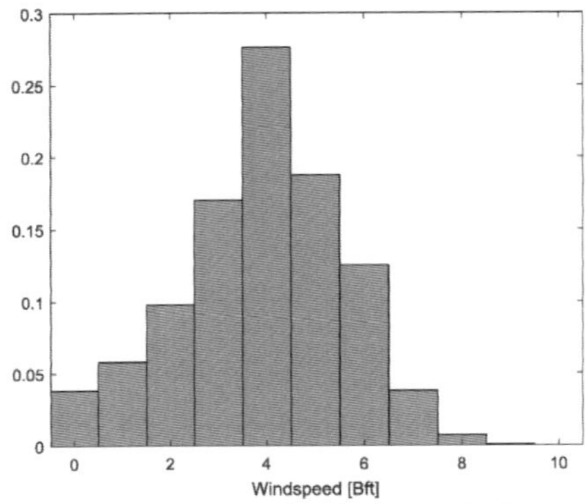

Figure 2: Histogram of wind speed for operational profile.

The wave scatter diagram in Figure 3 shows that waves up to 12 m do occur but in less than 0.05% of the time (denoted by a dash). Even in summer, there is a 0.1% chance of encountering a wave height between 8 and 9 m and a mean wave period between 12 and 14 s. This seems small, but it means the vessel could encounter these conditions on average 4 h annually (when no operational measures like rerouting are taken).

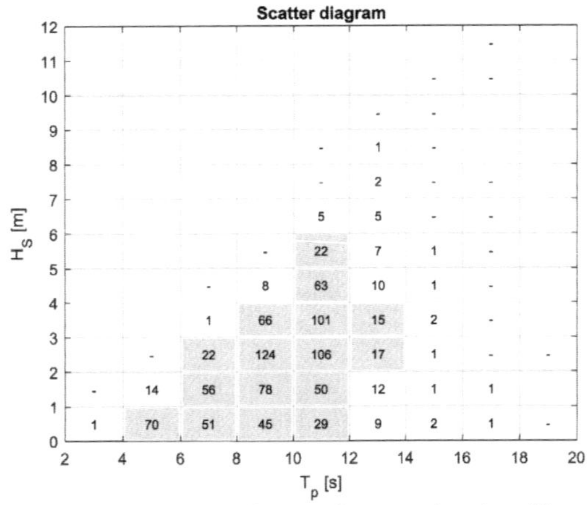

Figure 3: Wave scatter diagram for operational profile.

As the SRtP condition requires an assessment in Bft 8 it would be interesting to know what kind of wave conditions occur on this route. Figure 4 shows the resulting wave scatter diagram. It shows that the significant wave height could be between 1 and 11 m: the 1 m condition in the very early stages of a storm and the 11 m condition when a very heavy storm has just passed and the wind is decaying. The most frequently occurring wave condition is a significant wave height between 5 and 6 m and a peak wave period of 10 s and 12 s. The blue line in the plot is the Bft 8 line according to Janssen (see section 2.3). This shows that it correlates well with the Bft 8 wave scatter diagram on this route. Two wave conditions are selected for the case study, the most commonly chosen condition i.e. H_s=5m, T_p=13s (red open dot) and secondly H_s=5m, T_p=10.5s (red open square). The second condition is close to the most frequently occurring wave condition.

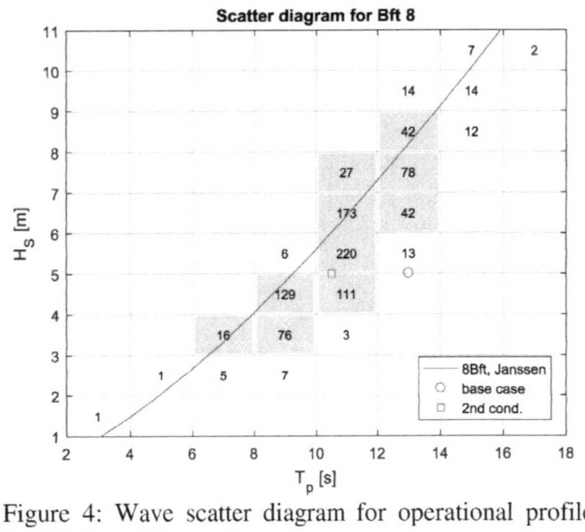

Figure 4: Wave scatter diagram for operational profile for Bft 8 condition only.

4.3 CALM WATER PERFORMANCE

From open water model tests, the propeller characteristics are obtained (see Figure 5). The wake fraction and thrust deduction fraction are obtained from the propulsion tests performed with both propellers active and from resistance tests (see Figure 6).

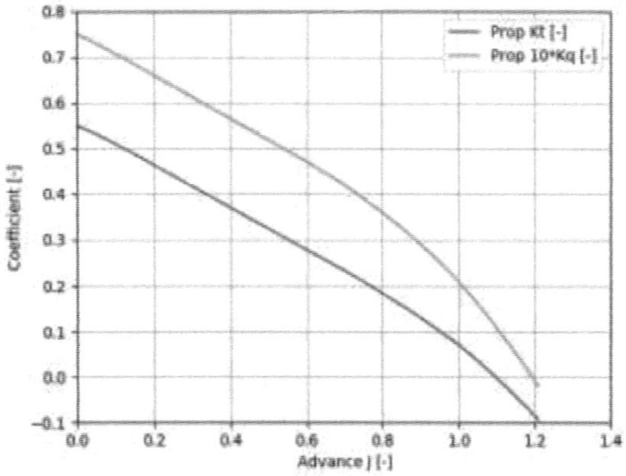

Figure 5: Propeller open water characteristics

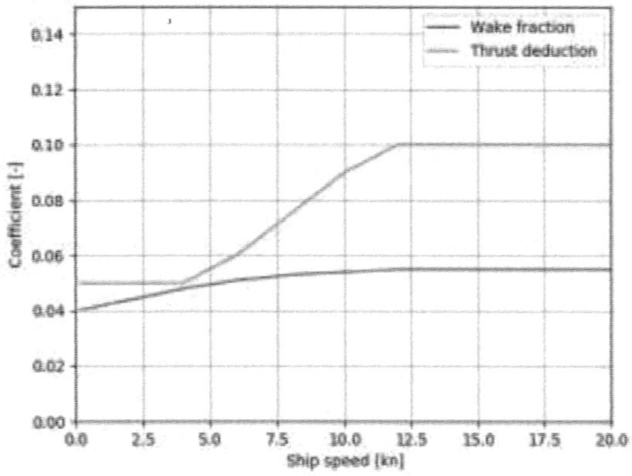

Figure 6: Thrust deduction factor and wake fraction as function of the ship speed

Finally, based on these results, the propulsion tests performed, and assuming an available shaft torque of 127.3 kNm, the cruise ship studied presents a design speed of 19.0 kn as seen from Figure 7 showing the available against required thrust.

If now we consider the case with one propeller blocked, the available thrust is divided by two and the required thrust to make speed is increased due to the presence of the blocked propeller. From Figure 8, one can see that in this situation the calm water speed already drops from 19.0 kn to about 12.8 kn.

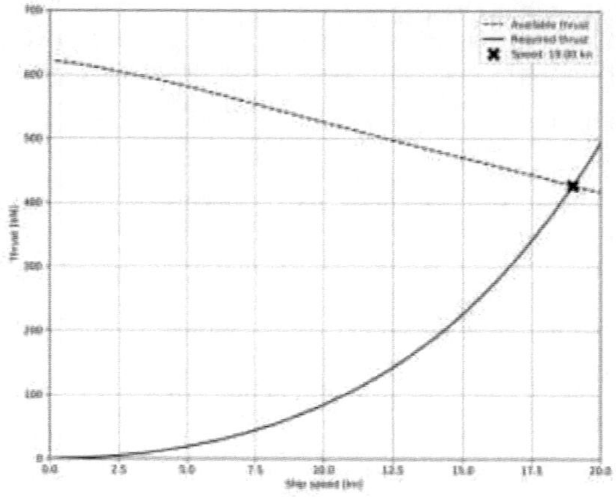

Figure 7: Calm water performance with both propellers working and onboard power equivalent to 19 kn design speed.

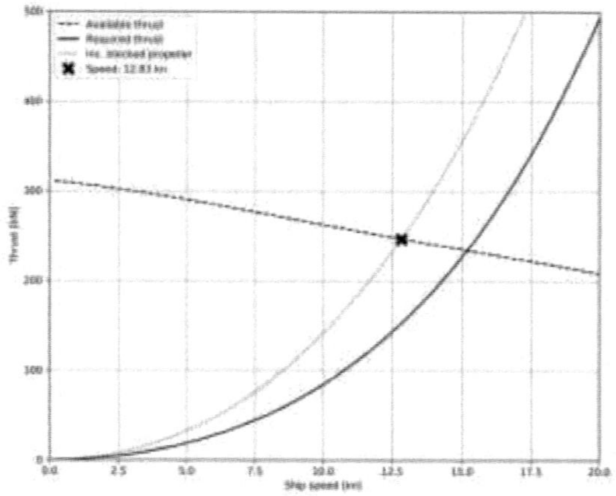

Figure 8: Calm water performance with one propeller blocked and onboard power equivalent to 19 kn design speed.

4.4 ADDED RESISTANCE DUE TO WIND AND WAVES

For the present case study, FATIMA computations were run at various speeds. In particular, Figure 9 presents the mean wave added resistance in head seas for a significant wave height of 5.0 m as function of the wave peak period for 3.0 kn, 6.0 kn and 9.0 kn.

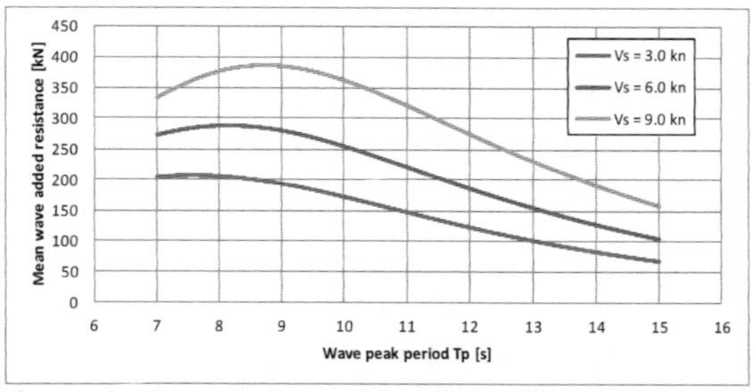

Figure 9: Wave added resistance in 5.0 m Pierson-Moskowitz head seas.

These results are based on a long crested Pierson-Moskowitz wave spectrum. From this figure, it is seen that at 6.0 kn for a peak wave period of 13.0 s, the ship wave added resistance is about 150 kN. In addition, this figure emphasises the fact that the wave added resistance is strongly dependent on the wave peak period. In particular, for the ship studied the wave added resistance goes from about 290 kN for 8 to 9 s wave peak period to about 100 kN for the longest wave period of 15 s.

In terms of wind added resistance, for the present ship a wind drag coefficient of 0.65 is considered together with a frontal area of 300 m^2.

4.5 SRtP ASSESSMENT – BASE CASE

By combining the above results, it is possible to calculate the sustained speed in 19.0 m/s wind and in head seas of 5.0 m significant wave height and 13.0 s peak period.

In particular, Figure 10 shows how much thrust is required to make speed due to the wind and wave added resistance in the case of a blocked propeller. From these results, one can see that the present ship design will be able to maintain a speed of 6.3 kn in the given environmental conditions. This is above the hard SRtP limit of 6.0 kn.

It is interesting to note that at the sustained speed, the calm water resistance is only 11% of the total (18% including the additional drag of the blocked propeller) and the wind added resistance makes a contribution of 22%. However, the largest contribution clearly comes from the waves: 60%. Although these contributions vary per ship, the wave added resistance dominates for all ships.

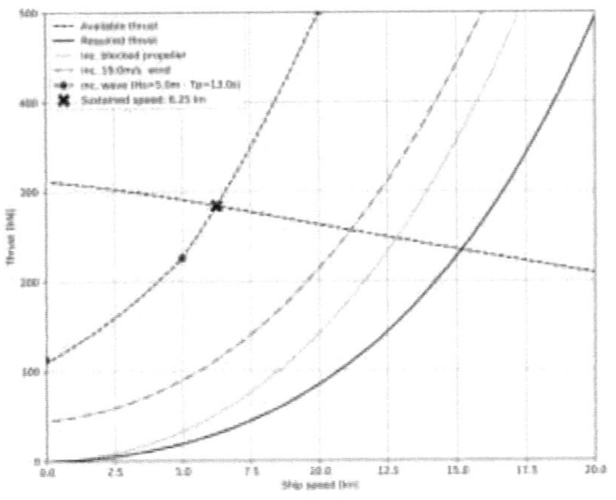

Figure 10: SRtP performance with onboard power equivalent to 19 kn design speed; in head 5.0 m, 13.0 s wave; 19 m/s wind and a blocked propeller.

The case of a perfectly water-milling propeller (i.e. no added thrust) is presented in Figure 11. From this figure it is seen that the sustained speed increases to about 6.7 kn in this more favourable condition.

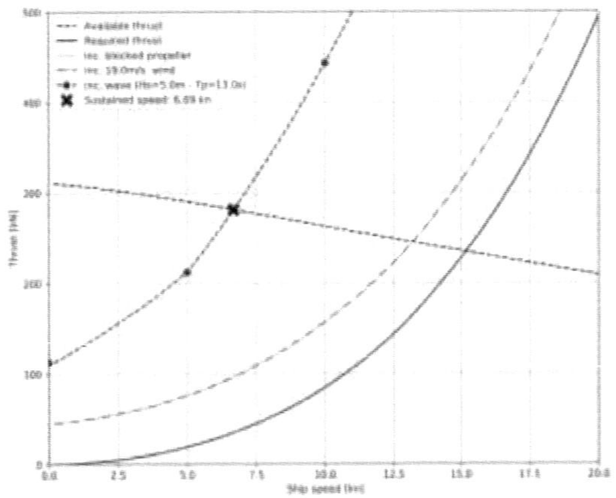

Figure 11: SRtP performance with onboard power equivalent to 19 kn design speed; in head 5.0 m, 13.0 s wave; 19 m/s wind and a water milling propeller.

4.6 SRtP ASSESSMENT – EFFECT OF ONBOARD POWER

So far, it was assumed that the ship had enough power to maintain in ideal calm water condition a design speed of 19.0 kn. This is actually quite fast for such a 120 m cruise ship. In this section, we now assume that the present cruise ship can maintain a sustained speed of 18.0 kn, which still leads to a Froude number of 0.27. Based on this reduced power, one can see from Figure 12 that the cruise ship can now only maintain 5.5 kn in the specified 19.0 m/s wind and head waves of 5.0 m significant wave height and 13.0 s peak period. This illustrative example shows how difficult it may be for a medium speed 100-150m cruise ship to maintain 6.0 kn in large head wind and head wave conditions.

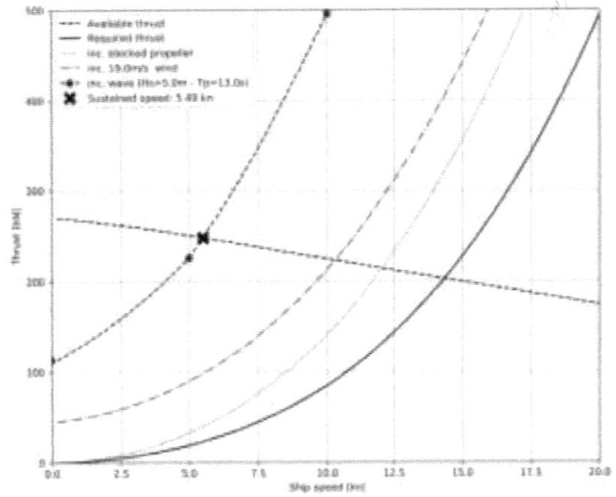

Figure 12: SRtP performance with onboard power equivalent to 18 kn design speed; in head 5.0 m, 13.0 s wave; 19 m/s wind and a blocked propeller.

4.7 SRtP ASSESSMENT – WAVE PEAK PERIOD

As discussed in Section 4.4, the wave added resistance is quite sensitive to the wave peak period. In particular, in the present section we investigate the SRtP performance of the studied cruise ship in 10.5 s wave period instead of the commonly chosen 13.0 s. This increases the wave added resistance by 30 to 40%. From Figure 13 it is seen that in this shorter sea state, the sustained speed drops from 6.3 kn to 4.7 kn. This shows that such weather condition can be very critical in case of actual engine failure.

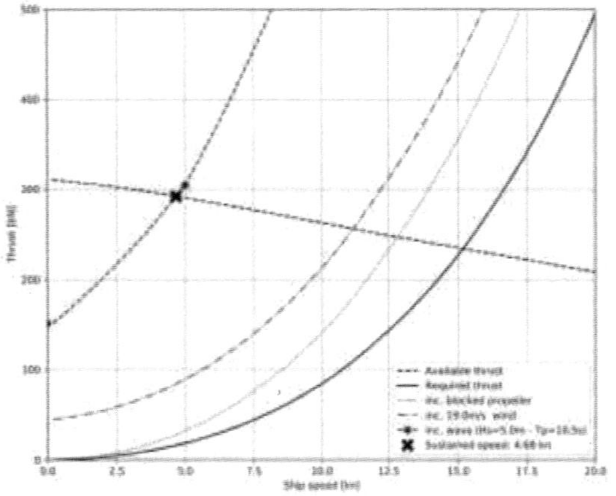

Figure 13: SRtP performance with onboard power equivalent to 19 kn design speed; in head 5.0 m, 10.5 s wave; 19 m/s wind and a blocked propeller.

4.8 SRtP ASSESSMENT – WAVE DIRECTION

Similarly to the wave period, one should not neglect the effect of the wave heading. A first step in accounting for the effect of wave direction is to consider the ship wave added resistance in bow quartering sea conditions. In particular, for the 100-200 m cruise ship the 13.0 s wave period might be more critical at 150 deg wave heading than in head seas. In the particular of the presently 120 m cruise ship, the wave added resistance increases by about 30-40% in 150 deg heading compared to head seas.

Figure 14 presents the SRtP performance for the studied cruise ship accounting for the increase of wave added resistance. From this figure, it is seen that in this sea state the ship's sustained speed drops to 5.7 kn.

Note that this example only accounts for the longitudinal mean force equilibrium. In quartering seas also the transverse wind and wave forces and the yawing drift moment should be in equilibrium. This typically results in a drift (and steering) angle, which also negatively affects the sustained speed.

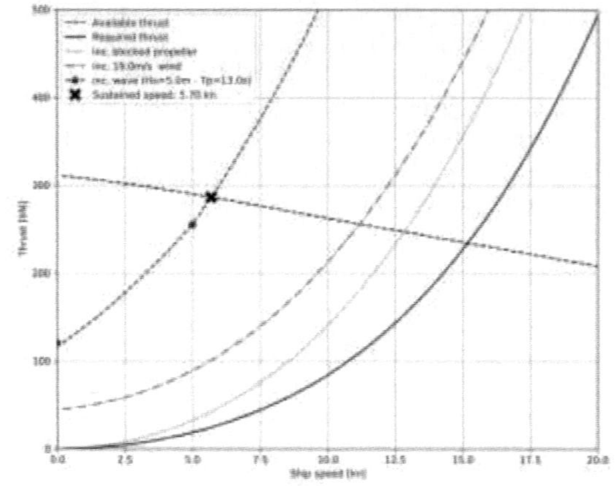

Figure 14: SRtP performance with onboard power equivalent to 19 kn design speed; in bow quartering (150 deg) 5.0 m, 13.0 s wave; 19 m/s wind and a blocked propeller.

5. CONCLUSIONS

The present paper discusses how to assess the compliance of cruise ships and RoPax ferries with the requirement for Safe Return to Port as detailed in SOLAS Ch. II-2/21.4.1 – Propulsion. Part of this requirement is the necessity to be able to maintain a minimum sustained speed of 6 knots in Bft 8 weather conditions for the worst single compartment damage case. This worst damage case is typically failure of one shaft line. From the methodology outlined and the presented case study, the following is observed:

- From the resistance and propulsion point of view, one of the key assumptions is whether the failed shaft line results in a water-milling or blocked propeller. In the present case study, the maximum sustained speed drops from 6.7 kn in the case of a water-milling propeller to 6.3 kn with a blocked propeller which is much closer to the required minimum speed.

- The design speed of passenger ships is a good indicator for compliance with the SRtP requirement of 6 kn. Ships with a design speed of around 18 kn or lower are often close to or below the 6 kn limit in case of one shaft failure and in 5 m waves. The 120 m cruise vessel is no exception. When the design speed is reduced from 19 kn to 18 kn, the sustained speed drops from 6.3 kn to 5.7 kn.

- In the SRtP requirements, the Beaufort number is clearly defined as being Bft 8. However, the wave conditions associated with this Bft 8 are not provided. As shown in this study the wave height as well as wave period could vary a lot at Bft 8. Typically a significant wave height around 5 m is selected. For the present case study, this seems to be a realistic condition, which occurs frequently. However, for different sailing areas this could be different. A typically applied peak wave period is 13 s in many studies. From the present work, it is shown that this is not a common wave period at Bft 8. A more typical peak wave period in these conditions is in the order of 10 s. Especially for smaller passenger ships this more frequently occurring wave period is rather demanding as it is at the peak of the wave added resistance. The case study shows that the sustained speed drops from 6.3 kn to only 4.7 kn when the peak wave period is decreased from 13 s to 10.5 s.

- Although the SRtP requirements require a head wind and waves condition, this is often not the most demanding condition. Typically both the wind and wave added resistance is larger in quartering seas. For the 120 m cruise vessel it is shown that when the heading is changed to bow quartering (150 deg), the sustained speed drops to 5.7 kn.

6. FURTHER WORK

It is shown that especially for relatively small and slow passenger ships it might be difficult to comply with the 6 kn minimum speed. It cannot be concluded only on the basis of this requirement that these vessels are unsafe. It might well be that although 6 kn is not reached, they are still in control. When the weather conditions improve, they could than safely return to port.

Time domain simulation could be performed to verify if the ship is able to maintain course in those Bft 8 conditions. Within the Cooperative Research Ships (http://www.crships.org), the ManWav time domain simulation tool was developed for manoeuvring in waves. This tool focuses on low-frequency wind and wave forces in combination with the ship's powering and manoeuvring performance to simulate low frequency track keeping, sustained speed and manoeuvring performances of the ship in different wind and wave conditions.

In order to verify the endurance, i.e. the time needed to reach a safe port, voyage simulations could contribute, as they simulate what would be the long term distributions of voyage time and fuel consumption properly including wind and wave hindcast data.

7. REFERENCES

1. Bunnik, T.H.J., 'Seakeeping calculations for ships, taking into account the non-linear steady waves', PhD thesis, Delft University of Technology, 1999.

2. Janssen, P.A.E.M., Komen, G.J. and Voogt, W.J.P., 'An Operational Coupled Hybrid Wave Prediction Model', Journal of Geophysical Research, Vol. 89, pp. 3635-3684, 1984.

8. AUTHOR BIOGRAPHIES

Rob Grin holds the position of Senior Project Manager at MARIN. He is responsible for amongst others seakeeping studies for cruise vessels and ferries. He is doing both scale model tests in the seakeeping basin and full numerical seakeeping studies. His specialisms are voyage simulations and wave-added resistance.

Rien de Meij holds the position of Senior Project Manager within the MARIN Cruise & Ferry market team. Besides this task, he is involved in hull form optimisation, numerical and experimental investigation of all types of ships.

Stephane Rapuc holds the position of Senior Project Manager at MARIN. He is responsible for amongst others seakeeping studies for cruise vessels and ferries. He is doing both scale model tests in the seakeeping basin and full numerical seakeeping studies.

CONCEPT OF DESIGN AND IMPLEMENTATION OF PILGRIM SHIPS IN INDIA

Yashwant R. Kamath, Final Year Graduate Student, Department of Ship Technology, Cochin University of Science and Technology, Kerala, India.
Dr. K. Sivaprasad, Professor, Department of Ship Technology, Cochin University of Science and Technology, Kerala, India.

SUMMARY

Being the land of Spirituality and Mythology, India has thousands of pilgrimage centres across its breadth and length. River Ganges, which originated from Great Himalaya and flows through number of states and finally merges at Kolkata with Bay of Bengal, is worshipped as goddess of prosperity by the masses in India. The goddess river Ganga has many pilgrimage centres spread along her shores right from the origin to the river mouth. The concept of Pilgrim Ships started in the year 1602, between Falmouth, England and Plymouth. From there has been more than 250 pilgrim ships plying mainly in Europe and America in the mid-17th Century. India, Specifically, the City of Bombay and parts of Gujrat became a part of global pilgrim ship route in the early 19th Century. There are also records of Pilgrim Ships built in India, during the reign of Mughal Emperor Akbar. All these has set forth a set of rules in the Indian Merchant Shipping act which gives specific rules for design of type vessel for pilgrimage. The proposed paper brings out design of a Pilgrimage Cum Cruise Vessel for Ganges, National Waterway No.1, with reference to the unique and specific design guidelines provided in the Indian Merchant Shipping Rules. The paper will discuss specific design issues related to requirements of pilgrims and general solutions to the requirements with a naval architectural touch. Potential for developing similar designs for other inland water-based routes in India will also be discussed as part of exploring a business proposal in this sector.

1. INTRODUCTION

Since the beginning of a practice of pilgrim ships in the year 1602, in England and its spread to India in the mid-17th Century, the fleet of pilgrim has grown and reached in peak in the mid-20th Century and faced a heavy decline after the advent of leisure tourism. The main route of pilgrim ships in India during those days were the Indian Muslims for their Hajj pilgrimage. Then, vast majority of Indians were unfamiliar, rather opposed the idea of quarantine, and opposed the strict sanitary measures on grounds of cost and unnecessary infringement into the privacy. But with all these, India faced international criticism because of its failure to provide adequate safeguards for the health of passengers on board pilgrim ships. Consequently, with reference to this, the Indian and British governments were strictly questioning the maritime sanitary policy. The European guests pointed out the lack of sanitary provisions and adequate safeguards for the health of pilgrims, to which even the wealthiest passengers were subjected to. Since then the provincial administrators began to compel the government to amend the then existing native Passenger Ships Act.

In 1870, The Passenger Ships Act was made stringent, but even then, the epidemics failed to seize especially during the Bubonic Plague and spread of Cholera in 1876. A system of Port Health officer was established in 1878 and Medical boards were enforced in 1882. Finally following the first Indian Merchant Shipping Act, 1923, an additional rule, named "Indian Pilgrim Ships rules" was introduced in 1933.

2. STATE OF THE ART RULES

The current applicable rules include the Indian merchant shipping act, the Indian Pilgrim Ship's act, and Inland Waterways Authority of India Act 1985.

2.1 INDIAN PILGRIM SHIPS RULES, 1933

This set of rules was formed following the uncontrolled decline of health and hygiene in Indian Maritime sector, especially in pilgrim ships. The Indian Pilgrim Ship Rules focusses mainly on the safety and living condition of pilgrims, the passengers. Currently, the Rules are issued by Director General of Shipping, under the Indian Merchant Shipping Act. Though most of the requirements in this rule will be met by the modern rules. The regulations under which affect the design and operation of vessel can be categorized as follows:

1. Life Saving Appliances
2. Fire Fighting Appliances
3. Fittings and accommodation
4. Supply of Food and water
5. Amount and Distribution of Baggage
6. Hospital Requirements

These broadly regulates the number of lifeboats/ life rafts, their standards and the equipment to be provided on-board. The number of spares are more compared to requirements of a general merchant ship. The rule mandates several drills to crew and the pilgrims on-board. The Indian Pilgrim Ship Rules specify the fire-fighting requirements of vessel based on the prime-mover of the vessel. The rules mandates the minimum menu of food items to be provided to different classes of passengers. There should be a provision to safeguard the costly belongings of the passengers upon their request. Also, since most of the pilgrim will be aged people, there should be adequate facilities for hospital requirements and also isolation facilities, in case of an epidemic like chicken pox.

2.2 INLAND WATERWAY AUTHORITY OF INDIA ACT 1985

This is the compilation brought out by the Inland Waterways authority of India (IWAI), which includes the then existing IWAI Act, Navigational regulations thereunder viz; (i) Prevention of Collision on National Waterways Regulations, 2002 (ii) National Waterways, safety of Navigation and Shipping Regulations, 2002 and (iii) Inland Waterways Authority of India (Classification of Inland Waterways in India) Regulations, 2006

This rule focusses mainly on the administrative, legal and contractual parts, Navigational requirements, like Light Shape and Sound signals, etc. And briefly on the lifesaving appliances.

3. THE CONCEPT SERVICE

The concept of service proposed in this paper is to run a "Pilgrim Cruise Vessel" to promote pilgrimage as well as tourism in India. The proposed "holy service" is intended to ply in major pilgrimage destinations along the Inland Waterway of India. The details of the service is discussed in detail below.

3.1 AREA OF OPERATION

The proposed area of operation is National Waterway 1 (N.W. 1). The 1620km stretch from Haldia to Allahabad. The N.W. 1 is currently under the focus of government for development in tourism. Curretnly there are 5 cruise vessels plying in these waters. More than thirteen to and fro cruises has been successfully completed the previous financial year as per the schedule. The main destination include Varanasi, Banaras, Prayagraj (Allahabad), Hooghly, Patna, Triveni, Haridwar.

The main festivals and rituals which attract the pilgrims are as follows:

Festival Name	Gangasagar Mela	Kumbh mela	Ganga Mahotsav	Ganga pooja	Ganga Dussehra
Observed on	January	January to march	January	March - April	April - June
Frequency	Annually	Once in 12 years	Annually	Annually	Annually
Location	Kolkata	4 different locations	Varanasi	Haridwar	Haridwar

Table 1. Festivals in N.W. 1

Festival Name	Dev Deepavali	Ganga Aarthi	Durga Pooja	Triveni Sangam	Shukrana Prakash Parv
Observed on	October to November	Daily	October	Daily	December to January
Frequency	Annually	Daily	Annually	Daily	Annually
Location	Varanasi	3 Different Locations	Haldia	Triveni	Patna

Table 1. (contd.)

3.2 MAIN PARTICULARS

The Main particulars of the proposed vessel are as follows:

Length	:	30m
Breadth	:	5m
Depth	:	1.8m
Propulsion	:	Electric Propulsion
Power Source	:	Solar + DG
Hull Material	:	Mild Steel
Complement	:	18PAX+Crew&Staff

The dimensions are selected so as to accommodate six number of 3-member Family, supporting staff and crew as per requirement and rules.

Required number of Crew and Staff are as follows:

Master	:	1 no.
Lascar	:	1 no.
Chef	:	1 no.
Travel Guide	:	1 no.
Priest	:	1 no.
Engine Crew/ Lifeboats-men	:	2 no.s

The vessel is intended to be run on solar power and on a Diesel Generator when there is no sufficient power available from the solar source.

3.3 SPECIAL FEATURES

To make the normal passenger vessel a Pilgrim Cruise Vessel following special features have been incorporated in the design. A concept general arrangement has been made and is shown below in .Figure 1.

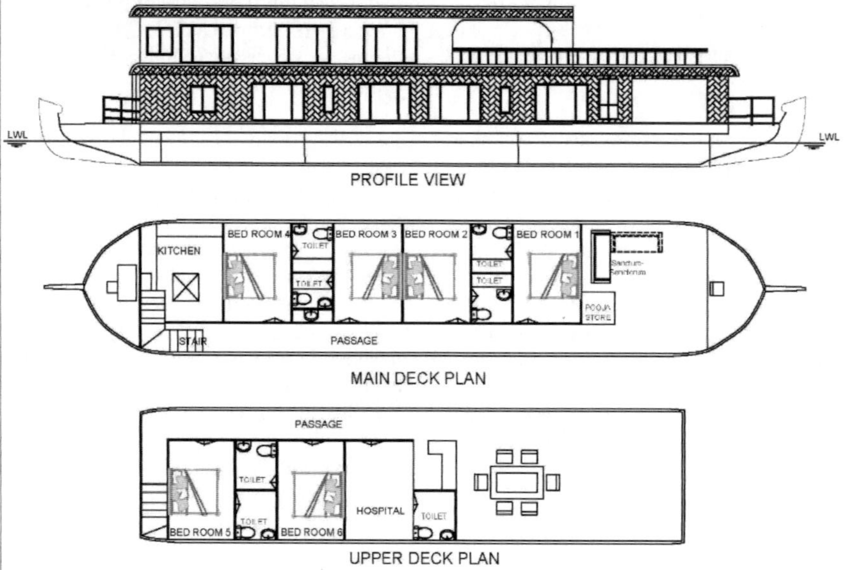

Figure 1. Concept G.A

3.3. (a) A Sanctum Sanctorum

A Sanctum Sanctorum, the holy place has been provided inside the vessel for day to day rituals of the pilgrim. According to the beliefs of the Hindu religion, the rituals are to be performed facing the East, since the vessels heading can't be always ensured, the Sanctum Sanctorum has been provided in such a way that, it can be moved, and rotated to face the right direction.

3.3. (b) Special Stores

Since the intended service of the vessel is to take the pilgrims to various temples along the N.W. 1, a special store has been arranged in the vessel to purchase items during the visit to temple or for their daily rituals.

3.3. (c) Baths

Common baths have been arranged to ensure purity to the pilgrims before entering the temple.

3.3. (d) Sun Deck

A sun deck with large floor area has been provided, for the pilgrims who believe in the sun god and perform rituals facing the sun or under sunlight.

3.3 (e) Additional Fire protection

There are large number of rituals where fire is involved or various purposes, all of which increase the risk of fire hazard. Therefore additional fire protection has been provided by providing fire boundaries to all accommodation cabins.

3.3 (f) Travel Guides

A travel guide will be available on-board to help the pilgrim know the places and regarding the practices in various temple.

3.3. (g) Medical Facilities

To provide health and fitness to the pilgrims, which mainly includes aged and disabled people, people who go in crowds increasing the probability of catching diseases. The vessel is also allocated with an isolation area in the open deck space, where any pilgrim can be isolated from others in case of an epidemic.

3.3 (h) Navigational features

Though the vessel is intended to be plying in the inland waters, modern navigational facilities of tracking and wireless telegraphy to be included in the vessel. Also, an additional gyro compass which is visible to the pilgrims will be provided on board for the pilgrims to get the direction.

3.4 REVENUE SAFEGUARD

It is clear from Table 1 that, there will be constant flow of pilgrims across the National waterway in different seasons of the year. But in the third quarter of every year there will be a steep decline in the passengers transported and also in other non-festival days in a year. In such conditions the revenue to the operating company can be ensured by using the vessel for leisure cruise in the N.W.

4. FUTURE EXPANSIONS

This current proposal can be extended to other Inland waterways in India. The vast network of Inland waterways and pilgrimage centers along the network can be thus exploited. Along with development of Pilgrim centers, the overall tourism sector is developed with such a service. An example for such an area of expansion will be the Pamba River in Kerala, South India.

5. CONCLUSION

The concept design of a Pilgrim Boat presented in this paper takes into account the salient features of inland waterway characteristics, the traditional, religious and ritualistic practices followed in India and the provisions of Indian Pilgrim Ship Rules, 1923.

The proposed design is capable of exploiting the vast potential that lies in the river based tourism. Though the attempt is restricted only to National Waterway-1, immense scope for similar application is readily visible in the other National Waterways and Inland Waterways of India and other state

This concept design would very well suite the requirements of mega waterway development plan such as *Namami Gange* (Salute the Sacred River- Ganga) implemented by the present Central Government.

6. AUTHOR'S BIOGRAPHY

Sivaprasad K. currently holds the position of Associate Professor at Department of Ship Technology, Cochin University of Science and Technology, Kerala, India. His area of specialization is Ship Production and Ship Repair and handles relevant classes for B.Tech Program and also Research Guide. He has published more than 35 papers to his name and has more than 20 years of teaching experience and worked as Senior Naval Architect at Mazagon Docks Ltd., India.

Yashwant R. Kamath currently a final year student of B.Tech in Naval Architecture and Shipbuilding at Department of Ship Technology, Cochin University of Science and Technology, Kerala, India. He has seven International papers to his name.

DESIGN FEATURES OF RIVER-SEA CRUISE PASSENGER VESSELS FOR RUSSIAN INLAND WATERWAYS AND ADJACENT SEAS

A G Egorov, **G V Egorov** and **I A Ilnitskiy,** Marine Engineering Bureau, Ukraine

SUMMARY

New concepts of river-sea cruise passenger vessels are not only new vessels, but also new tourist product. In market of sea and river cruises there are practically no offers on earlier popular cruises between ports of Black, Mediterranean and Baltic, and Caspian seas. Thus, absence of offers is connected, first, with lack of river-sea cruise vessels. For the solution of problem of connection of sea and river routes modern cruise vessels of PV09, PV300 and PV300VD projects developed by Marine Engineering Bureau are offered.
Researches have shown that generally for native conditions traditional single hull type passenger vessel can be recommended. CFD-modelling tests were executed to optimize hull form. Module principles of creating inner living space were used widely. Standard module cabins usage let alternate passenger capacity by using passenger cabin modules with different area in accordance with required comfort level.

1. INTRODUCTION

Today river and river-sea cruise vessels (CV) transport about 350-400 thousand tourists on Russian and Ukrainian rivers, which includes 75 thousand foreigners. Realities of native tourism are that to favourite river cruises of "entry" tourists from the USA, Germany, Australia on route Moscow – St. Petersburg, "partly" sea lines Moscow – Sochi, Kiev – Constanta, "circular" cruise on Caspian Sea were added.
At this moment in Russia, Ukraine and in general in world practice of operation of CV there is practically no fleet of river-sea cruise vessels.
From 254 river cruise passenger vessels built in Soviet period 98 (38.6%) have been written off (6.7% – 17 vessels were lost in accidents, 31.9% – 81 vessels were utilized). There were 32 vessels out of operation (12.6%) with middle age of 57.7 years. In operation – formally 124 vessels (48.8%) with middle age of 45.2 years. Mean age of utilization – 47.1 year. In the next decade this fleet most likely will be decommissioned and we will have only about 50 vessels [1] (without vessels which can be built during these years but which are not ordered yet).
By some expert estimates the market size will make 1.0 million tourists by 2030.
The market assessment including previous years experiment on operation of first-class vessels PV08 "Alexander Grin" and "Victoria" with native tourists allows to draw conclusion that, with taking into account the predicted reduction of vessels of old series, not less than eight new cruise vessels will be required by 2025.
Thus, problem of creation of river and river-sea cruise vessels became actual and important for native water transport and tourist industries.

2. AIM OF THE PAPER

Grounding on basis of 2010-2017 Marine Engineering Bureau (MEB) researches of concepts of modern river CV which are of interest for native shipping companies: PV300, PV300VD, PV09 projects. These vessels will be operated on classical river lines, and also make episodic cruises in coastal zones in accordance with assigned class.

3. MAIN TEXT

Operation of river CV on lines with sea parts is widespread in Ukraine. "Chervona Ruta" and "Viking River Cruises" cruise companies use river CV which earlier have worked along Black Sea coast [2].
The first cruises in Post-Soviet time on river CV in Black Sea became possible after executed by MEB work about expansion of sailing area of 302 project river passenger vessel of "T.G. Shevchenko" type. Work included grounding of assignment of river-sea "M-PR" class of Russian River Register (RRR).
Since then and till 2014 following vessels had been successfully operated both in Dnieper and Black Sea:
- two CV of 302 projects ("General Vatutin" and "Zvezda Dnepra" of 1988 year built) of "Chervona Ruta" cruise company;
- two CV of 301 projects ("Princessa Dnepra" of 1976 year built and "Mikhail Lomonosov" of 1979 year built of "Chervona Ruta" and "Viking River Cruises" cruise companies respectively).

In 2014 all Ukrainian river-sea lines were stopped. As for 2019 only two vessels are into operation on those routes: "Princessa Dnepra" of "Chervona Ruta" company and "Viking Sineus" (ex "Mikhail Lomonosov") of "Viking River Cruises" company of 301 projects. "Viking Sineus" was modernized in accordance with MEB PV17 project in 2013-2014.
After modernization it is four-deck, triple-screw motor vessel with overall length of 125.00 m, breadth of 16.70 m, with four-level superstructure along whole vessel's length, with ER located aft, with wheelhouse located fore, with double bottom, with inclined stem and cruiser aft end (see Figure 1). Passenger capacity is 196 passengers.
98 double-place cabins for location of 196 passengers are foreseen, including: 4 suites with balcony with area 30.1-31.6 m²; 2 "Deluxe" cabins with balconies with area

24.2 m²; 60 cabins with balconies with area 13.5-16 m²; 32 cabins with balconies with area 11.1-12.9 m².

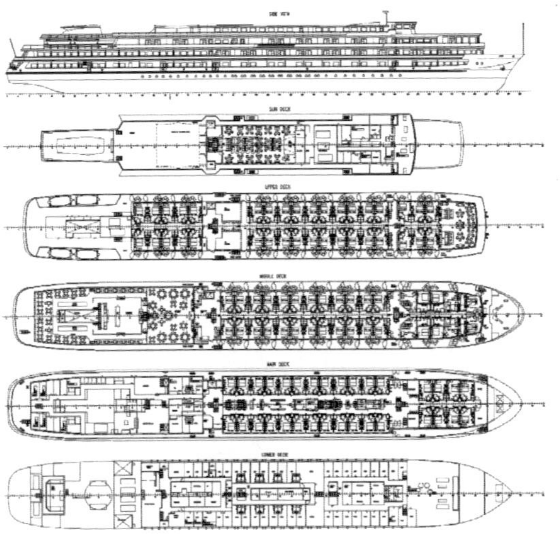

Figure 1. General arrangement of PV17 river-sea modernized cruise vessel

Passenger accommodations are located onto 3 decks: main deck, middle deck and boat deck.
The restaurant which allows feeding all the passengers in one shift is located onto the upper deck with area of 350 m² for 212 places.
Panoramic bar with area of about 100 m² for 55 places is in the fore part on the boat deck.
Music salon with bar with area of 150 m². for 100 places and large opened area with beach chairs and tents are located on the shade deck.
Lifesaving gear consists of four life boats and 46 inflatable freefall droppable rafts.

Vessel satisfies an increased ecological safety standard. Closed sewage and domestic water systems are foreseen. All sewage and domestic waters are forwarded to collector tanks. There is also sewage water treatment plant.
The next example of river-sea CV is Russian "Rus' Velikaya" vessel. She was built in 2013 with usage of elements of existing vessel of 588 project "N.F. Vatutin". Vessel's length – 95.6 m, breadth – 13.5 m, draught – 2.45 m. Passengers' capacity – 196 persons, crew – 66 persons.
But all these vessels are modernized.

4. FEATURES OF NEW RIVER AND RIVER-SEA CRUISE PASSENGER VESSELS

To our opinion, new generation of cruise vessels for Russian inland waterways will be characterized by the following features.
Due to main dimensions. Restrictions of way conditions of vessel's operational region are determinative (e.g. depths, dimensions of vessel's navigation pass, dimensions of hydraulic engineering facilities, etc.) – "Volga-Balt Max", "Volga-Don Max" / "Dnipro Max", "Belomor-Baltiysk (BBK) Max", "Danube Max" classes. Main dimensions of CV concepts are accepted depending on data, which are shown in Table 1. For further details see [3]. It is clear that river-sea concept will have restrictions on passengers' capacity as length and breadth of the vessel appointed in accordance with way conditions will limit useful area which can be allocated for cabins and for passenger serving spaces (restaurants, salons etc.). The number of decks is defined by air draught.

Table 1. General information about possible main particulars of river-sea CV

Vessel's class		Overall length L_M, m	Overall breadth D_M, m	Draught d, m	Air draught H_{AD}, m
"Volga-Balt Max"		≤ 185	≤ 16.95-20.1	2.90-3.80	≤ 13.2
"Volga-Don Max" / "Dnipro Max"		≤ 140	≤ 17.0	3.20-3.70	≤ 13.2
"BBK Max"		≤ 132	≤ 13.63	2.90-3.80	≤ 11.9
"Danube Max"	Danube – Passau	≤ 135	≤ 13.40	2.00-2.50	≤ 6.7
	Upstream Passau*	≤ 135	≤ 11.45	1.70-2.00	≤ 6.03
* – including pass-through to Northern Sea (Danube-Main-Rhine system)					

Due to architectural and constructive vessel's type, as follows [4]:
- so called "floating hotels" with superstructure at all vessel's breadth without passages by sides (with full-grown balconies / terraces or with French balconies at cabins for CV that are oriented to foreign tourists. Upper "sun" deck is equipped with swimming pool and tents;
- with promenade gallery by sides on all decks; such arrangement provides all-round view for tourists. This variant is used rarely for new projects because many tourists don't like when other people are close to the windows of their cabins;
- with fore saloon that provides both-sides view;
- interim option (for CV oriented on Russian tourists) is applied. This option combines variants mentioned above in different degrees. Practically all new CV (sea and river) built abroad are foreseen with maximal number of balconies.

Due to inner compartments planning:
- module principles of forming living blocks in total and cabins themselves (standard cabins) and vertical zoning of living and public compartments were used;
- restaurant, salons and bars are arranged in vertical zones separate from living compartments;
- noise spaces are not arranged above passenger or crew cabins. Total cabins area is 16-43 m² including balcony / French balcony (cabins are double usually);

- restaurants with total places equivalent to passengers' number, conference hall, bars, internet-salon, baby rooms, beauty salon, souvenir shop, fitness salon, solarium are located on the open sun deck;
- lifts connected all decks, communications and toilets are foreseen for physically challenged people.

It is unequivocally that in process of design of new vessels one should start from the comfort level set by the customer as a quantity of conditional stars or comfort factor which will serve as the regulating factor of the future vessel's efficiency as function that determines required areas of decks and compartments, overall vessel's dimensions, vessel's propulsive characteristic, and vessel's economic efficiency as a result. As a criterion of an estimation of passenger vessels design comfortableness it is possible to apply the five-stars scale informally used practically by all tour operators. The quantity of stars increases according to increase of service quality and comfort. It is necessary to emphasize especially CV with enhanced comfort (per se, yachts) and vessels for single-day cruises (cabinless). Such vessels often don't relate to CV due to their significant differences in the approaches to passenger zone and several special questions (availability of additional opportunities for water types of activity, etc.). However, such vessels are often become platform for development including for classical cruise transportations. Example of this thesis is shown on Figure 2 (comparison of side views of PV09 and PV300, PV300VD projects). Already realized in 2017 PV09 river-sea cruise passenger vessel became the first in "line-up" of new generation cruise vessels (see Figure 3).

Choice of classification society class of perspective cruise passenger vessel assumed following: assignment of class based on sailing areas in accordance with planned directions of transportations; definition of ice class in accordance with saved-up experience and tendencies to extension of the navigation period including winter operation; assignment of economically reasonable life-time cycle of the vessel.

For safe and accelerated process of passengers' evacuation modern evacuation systems consisting of sleeve and accepting platform (sometimes landing of passengers can be carried out straight on rescue device) on which passengers during evacuation go down and with which passengers move on rescue vessel boats and rafts can be used.

Crew must be formed with considering ship power plant automation, automation of deck works (automatic anchor mooring winches), automation of process of loading of supplies and mechanization of process of cleaning of decks (washing machines, etc.) and with taking into account usage of specialties combination.

Choice of main particulars of river and river-sea CV is defined by way conditions, such as tour operator strategy and his position in the market, adherence to those or other lines and comfort level in view of necessities, propensities and interests of passengers for which these vessels are built. Such external subjective choice should be carried out basing on discrete number of the alternatives which quite objectively have arisen in the domestic tourist market as a result of the compromise due to choose of way restrictions and sizes of tourist groups (400-500 people, 250-300 people, 200-250 people, 100-150 people).

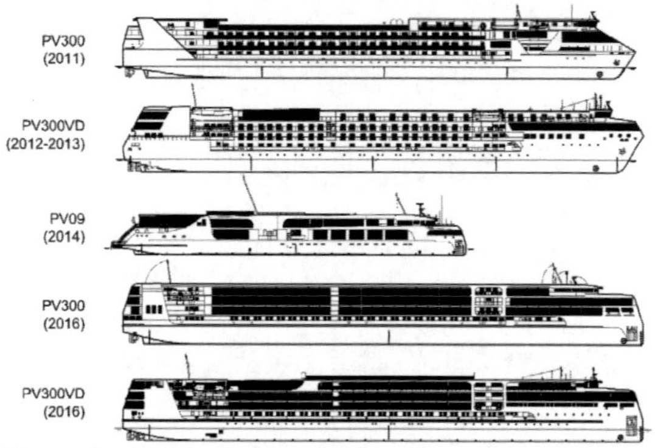

Figure 2. Comparison of side views of PV09 and PV300 / PV300VD concepts

Analysis of such alternatives has allowed to build parametrical line-up of river and river-sea cruise vessels objectively demanded by native Ship-owners and on its basis to develop in MEB projects of new river-sea vessels.

The most attractive is "Volga-Don Max" / "Dnipro Max" PV300, PV300VD concepts (dimensions are defined by way conditions of Volga-Don Ship canal, Dnipro River, passengers' capacity is about 300 people). Such vessels are planned to be operated on lines connecting St. Petersburg port and Moscow port, on Volga-Don Ship Canal with exit in Azov and Black Seas, on Volga with exit in Caspian Sea, on Volga-Baltic waterway with exit in Baltic Sea or on Kiev – Kherson – Nikolayev – Odessa – Izmail – Constanta line.

5. PV300VD CONCEPT

PV300VD concept architectural-structural type is as follows: steel, triple-screw motor ship, with slopping stem and transom stern, with excess freeboard, with extended forecastle, with four-tiers living superstructures through the whole vessel's length located amidship, with wheelhouse located fore and engine-room located aft (see Figure 4). Comfort level on PV300VD vessel will correspond 4-5* hotel standard.

PV300VD "Volga-Don Max" class vessel will work on lines Moscow – St. Petersburg and will make voyages Moscow – Sochi, Astrakhan – Caspian ports.

The vessel's class will allow also to visit ports at Black and Mediterranean seas.

General arrangement of PV300VD vessel is presented on Figure 5. Vessel's main are as follows: length overall 141 m; length between perpendiculars 135.56 m; breadth overall 16.82 m; breadth 16.60 m; depth 5.50 m; draught maximal 3.20 m; air draught from BL 16.25 m; crew and service staff 90 people; autonomy by fuel 15 days; autonomy by other stores 10 days.

Figure 3. General view of river-sea cruise vessel of PV09 project

Passenger capacity of the vessel can be varied from 300 to 400 depending on the type of cabin modules used (area of cabins is 16-43 m^2) and usage of additional places for children.

Figure 4. General view of PV300VD river-sea cruise vessel

Arrangement of the cruise vessel in the variant of the Moscow River Shipping Company (shipowner of PV300VD vessel) foresees 310 passengers who travel in 155 comfortable cabins of different classes: 7 luxury apartments with a balcony of about 36-43 m^2 area (33-35 m^2 without considering balcony); 3 luxury cabins with a balcony of about 26 m^2 area (21 m^2 without considering balcony); 2 luxury cabins for disable people with a balcony of about 33 m^2 area (27 m^2 without considering balcony); 133 standard cabins with a balcony of about 19-21 m^2 area (16-18 m^2 without considering balcony), see Figure 6; 8 standard cabins of about 16 m^2 area; 2 standard cabins for disable people of about 21 m^2 area.

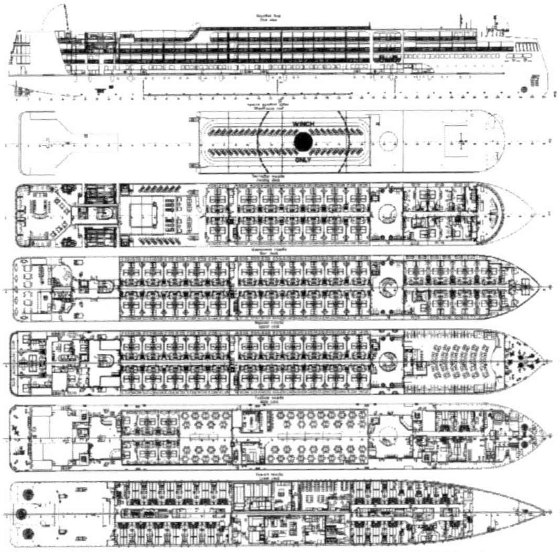

Figure 5. General arrangement of PV300VD river-sea new generation cruise vessel

Passenger cabins are in the middle vertical zone on 4 superstructure decks. Cabins on three upper decks have own balconies with sliding sections that provide through deck pass at emergency situations.

All passenger cabins are equipped with double beds, lavatory and douche, air conditioner, wardrobe, safe, TV set with satellite and internal vessel's channels, hair-dryer, sockets 220V and 110V, wireless Internet (Wi-Fi), vessel's radio translation, external and internal telephone connection.

Figure 6. Design view of standard passenger cabin on PV300VD river-sea cruise vessel

Fore, middle and aft vertical zones are used for providing rest of passengers on the vessel. Forward vertical zone includes as follows: 380 m^2 panoramic theatrical music saloon with a bar, gift shop, hairdressing saloon, a spa saloon with sauna, fitness-centre, massage room and rest room with jet tub. Middle vertical zone includes as follows: double level solarium with grill-bar, 530 m^2 main restaurant. Aft vertical zone includes as follows:

170 m² 24 days bar/disco, restaurant with 110 m² opened kitchen.

One change meal period for passengers is foreseen on the vessel. Two restaurants are applied for this purpose: main restaurant and restaurant with opened kitchen.

The main restaurant has two halls (of 360 m² and 140 m²) equipped with smorgasbord. The bigger hall is equipped also by bar console.

Meals preparation can be watched in the restaurant with open kitchen because kitchen is separated from restaurant by glass wall only.

All restaurants have maximal available glass cover that enables to take pleasure not only in tasty meal but also in window's view.

Bar is worked till the last client. Here is bar in the daytime and disco in the night. Bar's arrangement does not prevent to have a rest to other passengers. Double-tier sun deck is located near; there are grill bar, swimming pool and 4 Jacuzzis. Part of deck can be covered by moving tent when weather is bad.

PV300VD is not only up-to-date high comfortable and handy vessel but also safe, ecology "clear" passenger one that fully meets requirements of all International Conventions (IC), including IC for the Safety of Life at Sea (SOLAS), IC for the Prevention of Pollution from Ships (MARPOL), International Regulations for Preventing Collisions at Sea (COLREGS). Additionally, the vessel meets all Russian requirements such as Sanitary Regulations and Norms, Fire Safety Regulations, Technical Regulations for Safety of Inland Water transport Objects.

The vessel fully satisfies SOLAS norms and Class requirements of damage stability, including probability assessment of hull subdivision, special requirements for passenger vessels of Russian Maritime Register of Shipping, and all normative requirements of Russian River Register Rules.

Main 7 watertight transverse bulkheads are installed on the vessel. They divide hull into 8 impermeable compartments. At the aft part of the vessel from each side two fast 150 persons' capacity life-boats and two workboats are installed. For safe and accelerated process of passengers' evacuation modern evacuation systems consisting of sleeve and accepting platform (sometimes landing of passengers can be carried out straight on rescue device) on which passengers during evacuation go down and with which passengers move on rescue vessel boats and rafts can be used. 2 such evacuation systems with closed inflatable self-righting rafts (4 rafts on 101 persons each) with fore anteroom are installed on vessel. Earlier such systems on river and river sea vessels were not applied.

Middle part of vessel's hull is constructed by mixed framing system (see Figure 7). Main and upper decks are constructed by longitudinal framing system on whole length. Lower, boat and tent decks are constructed by transverse framing system on whole length. Bottom and double bottom are constructed by longitudinal framing system in middle part and transverse framing system in ends and in ER. Outer side is constructed by transverse framing system on whole length. Constructions of ends are constructed by transverse framing system.

Construction of sides and bottom is reinforced in accordance with RRR rules and ice class "Ice 30".

For ensuring of maximal glazing execution of cuts for windows and doors practically from deck to deck required solution of number of problems connected with strength of attachment fittings and material of windows.

Construction of vessel's decks in area of big cutaways under atriums is reinforced with thickened sheets and additional connections.

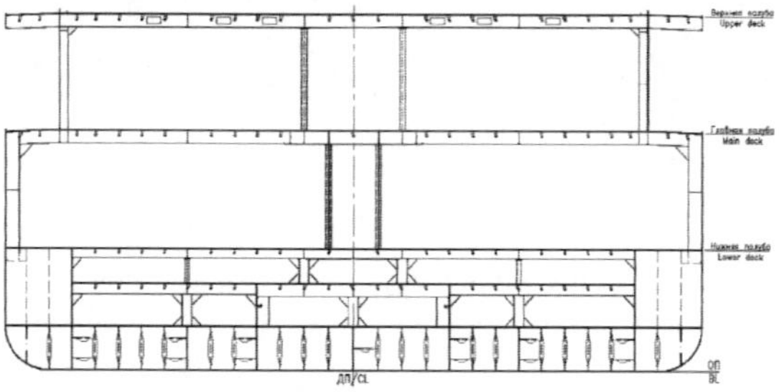

Figure 7. Transverse cross-sectional view of PV300VD vessel's hull

Fire safety of new concept is provided by division of the vessel into main vertical zones with constructive and thermal barriers; subdivision of accommodation spaces from other spaces with thermal and constructive barriers; exception of application of inflammable materials; detection of any seat of fire in zone of its emergence; restriction of distribution and fire extinction in places of its emergence; protection of means of escape and accesses to them for fight against the fire and readiness of fire-protection equipment to fast application.

From each waterproof compartment located below bulkhead deck two outmost from each other exits are made.

Special requirements to equipment of ventilation systems of the vessel were revealed on design stage. Spaces which are in different vertical fire zones are not connected between each other through vent ducts. Vent ducts are arranged in way so that they could be blocked from the outside of ventilated spaces.

Ventilation control is executed from two posts located on maximal outmost from each other.

Except meaningful measures of fire protection, complex of actions for fit out of accommodation, public and service spaces by automatic systems of fire detection and automatic sprinkler systems in local vertical fire zones is executed.

In aft part of the vessel three full-rotate fixed pitch rudder propellers (FPRP) with mechanical power on input shaft is of 1000 kW each are installed. For improvement of controllability at low speeds, at passing narrow waters and at moorings bow thruster is foreseen.

Traditionally three-shaft unit is used on big Russian river CV as propulsion complex. PV300VD concept is no exception. The sense of such decision absolutely clear: if it is necessary to go with full speed three screws are working, with operational speed is of 20 km/h – two

screws, with speed is of 12 km/h (maximally possible speed for many parts of IWW in accordance with Rules of Sailing on Russian IWW, f.e. Canal named for Moscow, Svir river etc.) – one screw.
CFD-modelling tests were executed to optimize hull form (underwater part of hull is the same on PV300 and PV300VD projects), see Figure 8 and 9. Researches have shown that traditional single hull type passenger vessel can be recommended. According to received results of calculation of PV300VD concept full speed makes 26.3 km/h (14.2 knots) with 3 main engines 1000 kW each (on 85% of maximal continuous rating).

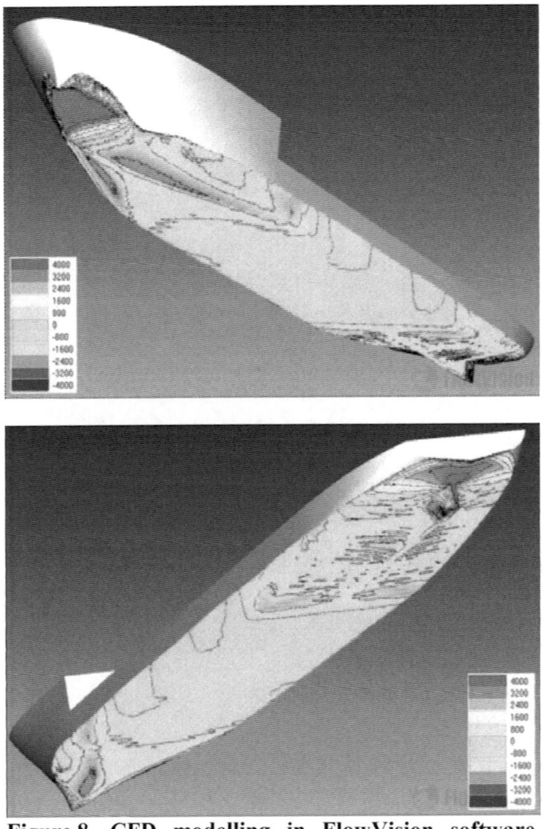

Figure 8. CFD modelling in FlowVision software. Pressure distribution. Fore and aft view

Usage of multi-shaft unit is also caused by aspiration to improve manoeuvrability of the vessel and to increase propulsion performance factor at limited water propeller diameter due to distribution of power between three propulsion units and corresponding decrease of their loading on thrust.
For providing full control over the vessel's movement while manoeuvring on slow speed, mooring operations and docking installation of vessel's control system which unites all propulsion units of the vessel (screws, FPRPs and bow thrusters) in integral module is provided and thus control of all these propulsion units is executed by means of one joystick.
Main power plant is diesel-electric. It consists of four main diesel-generators of alternative current 690V, 50 Hz with electrical power 1140 ekW each, feeding electric propulsion plant and other customers.

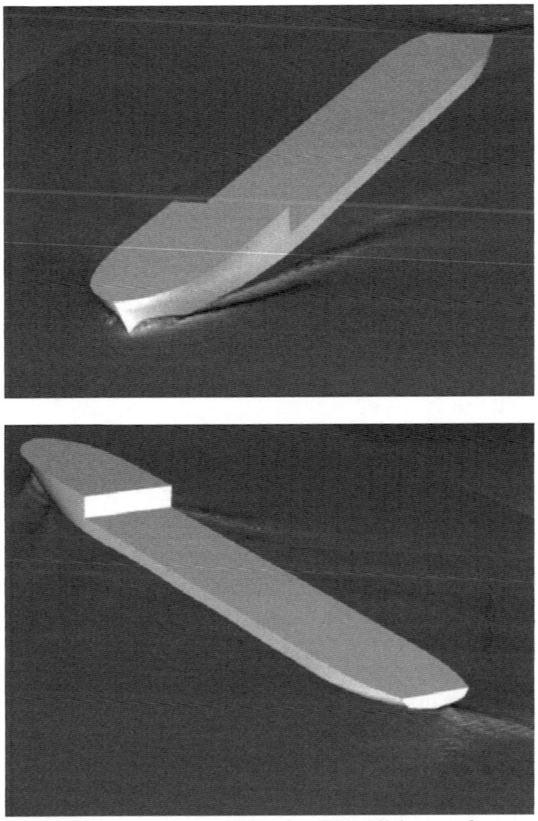

Figure 9. CFD modelling in FlowVision software. Free surface. Fore and aft view

For cruise passenger vessel, for which needs of passenger part for energy are commensurable with needs for energy for the movement, use of electric propulsion provides considerable advantages on building stage and, especially, in operation. Electric propulsion has allowed to exclude on building stage one diesel-generator on power and to reduce in general number of diesels – 4 main diesel-generators are installed instead of 3 main engines and 3 diesel-generators. During standard operation two main diesel-generators are providing power to propulsion motors and other vessel consumers. The third diesel-generator is in reserve or works when speed is increased or under bad weather conditions. And it is possible to carry out technical maintenance on the fourth diesel-generator. Diesel-electric SPP provides higher profitability on low and half speeds, and also during vessel voyages with frequent stops and manoeuvres. Electric drive allows to apply screw-propellers with the best performance coefficient and to use primary engines in the most economy mode because between primary engine and screw practically any gear ratio can be chosen.
Auxiliary power plant consists of: emergency diesel-generator of 214 kW electrical power; auxiliary boiler plant consisting of two oil fired thermal oil heaters with

heat power 800 kW each, fore thermal oil exhaust gas heaters with heat power 250 kW each; auxiliary mechanisms and apparatus working for power plant.
As fuel for main engines IFO380 with viscosity 380 cSt at 50°C are used. For emergency diesel-generator MDO with flash point over 60°C are used.
6 persons passenger lifts for moving passengers between decks are installed (1 in fore hall and 1 in aft hall).
2 cargo lifts for moving cargoes between galley and restaurant are foreseen. Cargo lift from laundry complex on lower deck connecting all decks of the vessel is provided.
Master's cabin is located on the tent deck near the wheelhouse. Accommodation of other 89 members of crew and service stuff is foreseen in fore vertical zone and in hull in middle vertical zone. All necessary rooms required by sanitary regulations are foreseen for crew and the service staff:
Crew consists of 10 persons of command squad (including cruise director and doctor), 10 persons of deck and engine team and 70 persons of service stuff (hotel and restaurant).

6. PV300 CONCEPT

PV300 "Volga-Don Max" class vessel will work on lines Moscow – St. Petersburg, and also will make voyages in coastal zones in accordance with assigned RRR class ✠ M-PR 3,0.
PV300 concept architectural-structural type is as follows: steel, self-propelled river-sea motor ship with two full-rotated rudder propellers, with plumb stem and transom stern, with excess freeboard, with extended forecastle, with four-tiers living super-structures through the whole vessel's length, with wheelhouse located fore and engine-room located aft. Comfort level on PV300 vessel will correspond 4-5* hotel standard.
General arrangement of PV300 vessel is presented on Figure 10. Vessel's main particulars are as follows: length overall 141 m; length between perpendiculars 135.36 m; breadth overall 16.80 m; breadth 16.60 m; depth 5.00 m; draught maximal 3.00 m; air draught from BL 16.25 m; crew and service staff 144 people; autonomy by fuel 15 days; autonomy by other stores 5 days.
Arrangement of the cruise vessel in the basic variant of "Vodohod" shipping company (shipowner of the vessel) foresees 342 passengers who travel in 171 comfortable cabins of different classes: 1 luxury cabin of about 29 m² area; 1 luxury cabin of about 26 m² area; 168 standard cabins of about 19 m² area; 1 standard cabin for disabled people of about 29 m² area.
Passenger cabins are located in the middle vertical zone on 3 superstructure decks.
All passenger cabins are equipped with double beds, lavatory and douche, air-conditioner, wardrobe, safe, TV set with satellite and internal vessel's channels, hair-dryer, sockets 220V and 110V, wireless Internet (Wi-Fi), vessel's radio translation, external and internal telephone connection.
Fore, middle and aft vertical zones are used for providing rest of passengers on the vessel. On sun deck rest area for taking of solar baths and run track are foreseen.
On main deck rest areas with passenger lifts and reception desks are foreseen in fore and aft lobbies (atriums).
Souvenir shop is foreseen in fore lobby on main deck.
PV300 as PV300VD vessel also safe, ecology "clear" passenger vessel that fully meets requirements of all Russian requirements such as Sanitary Regulations and Norms, Fire Safety Regulations, Technical Regulations for Safety of Inland Water transport Objects. The vessel fully satisfies all normative requirements of Russian River Register Rules for passenger vessels.
Main 7 watertight transverse bulkheads are installed on the vessel. They divide hull into 8 impermeable compartments.

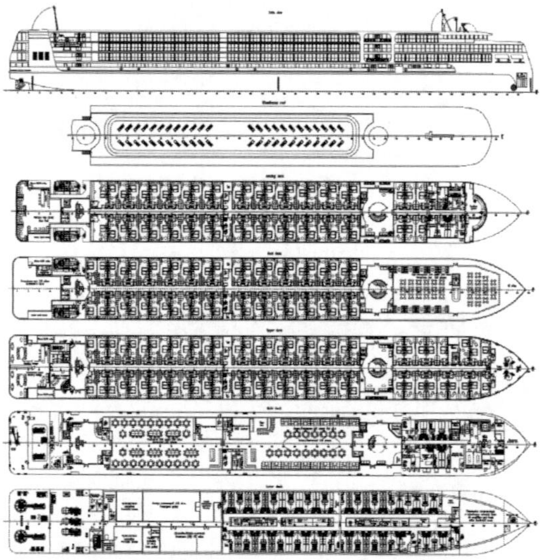

Figure 10. General arrangement of PV300 river-sea new generation cruise vessel

At the aft part of the vessel from each side two fast 48 persons' capacity life-boats and two workboats are installed. For safe and accelerated process of passengers' evacuation 2 evacuation systems with closed inflatable self-righting rafts (2 rafts on 100 persons each) with fore ante-room are installed on vessel.
Middle part of vessel's hull is constructed by mixed framing system. Main and upper decks are constructed by longitudinal framing system on whole length. Lower, boat and tent decks are constructed by transverse framing system on whole length. Bottom and double bottom are constructed by longitudinal framing system in middle part and transverse framing system in ends and in ER. Outer side is constructed by transverse framing system on whole length. Constructions of ends are constructed by transverse framing system.
Construction of sides and bottom is reinforced in accordance with RRR rules and ice class "Ice 30".

Construction of vessel's decks in area of big cutaways under atriums is reinforced with thickened sheets and additional connections.

Fire safety of new concept is provided by division of the vessel into main vertical zones with constructive and thermal barriers; subdivision of accommodation spaces from other spaces with thermal and constructive barriers; exception of application of inflammable materials; detection of any seat of fire in zone of its emergence; restriction of distribution and fire extinction in places of its emergence; protection of means of escape and accesses to them for fight against the fire and readiness of fire-protection equipment to fast application.

From each waterproof compartment located below bulkhead deck two outmost from each other exits are made.

Special requirements to equipment of ventilation systems of the vessel were revealed on design stage. Spaces which are in different vertical fire zones are not connected between each other through vent ducts. Vent ducts are arranged in way so that they could be blocked from the outside of ventilated spaces.

Ventilation control is executed from two posts located on maximal outmost from each other.

Except meaningful measures of fire protection, complex of actions for fit out of accommodation, public and service spaces by automatic systems of fire detection and automatic sprinkler systems in local vertical fire zones is executed.

In aft part of the vessel two FPRP with mechanical power on input shaft is of 1200 kW each are installed. For improvement of controllability at low speeds, at passing narrow waters and at moorings bow thruster is foreseen.

As special researches have shown better manoeuvrability of the vessel is provided on wide angles at FPRP turn. In process of decrease in motion speed of the vessel advantage of FPRP increases up to the lowest speed <3 kn when the vessel with rudders becomes almost uncontrolled and FPRP allow to carry out vessel turn in pos.

Usage of FPRP is especially effective while vessel's operation in river when passing narrow places and numerous rivers turns the vessel cannot move with maximum speed and efficiency of rudders falls down. Thus, the vessel equipped with FPRP can pass the same way section faster.

If to operate vessel on one FPRP there is no loss of controllability and thus vessel speed falls slightly to 7.5-8.0 kn. In economy mode of operation on one FPRP considerable fuel saving exists.

Purchase value of FPRP is approximately equal to the total cost of speed-reduction gear, shaft line, stern gear, screw, nozzle, steering gear and rudder.

Therewith FPRP installation is simpler: there is no need for laying of shaft line, thus less qualified personnel can be used, time of FPRP installation is less than time of mechanical transmission installation. All this allows to cut down expenses on mounting.

FPRP installation can be carried out afloat after vessel's launching. Delivery and installation of equipment of rudder propeller system can be made for some months later that also leads to economy of funds.

Therefore, FPRPs have been applied on PV300 concept.

Main power plant is diesel. It consists of two main diesel engines Wartsila 6L20 of maximal continuous power 1200 kW each.

Auxiliary power plant consists of: three diesel-generators of electrical power of 760 ekW each; emergency diesel-generator of 214 kW electrical power; auxiliary mechanisms and apparatus working for power plant.

Auxiliary boiler plant consists of one oil fired thermal oil heater with heat power 800 kW; three thermal oil exhaust gas heaters with heat power 170 kW each.

As fuel for main engines IFO380 with viscosity 380 cSt at 50°C and low sulphur content is used. For emergency diesel-generator MDO with low sulphur content with flash point over 60°C is used.

7. CONCLUSIONS

The most serious restraining factor of growth of native cruise tourism industry is lack of modern river and river-sea cruise vessels.

Mean age of such vessels makes 45 years, thus age of half of 90 existing CVs is 50-60 years. In the next decade this fleet most likely will be written off and we will have only about 50 vessels.

For the solution of problem of updating it is offered to connect river and some coastal sea routes on the basis of modern cruise vessels of PV300, PV300VD, PV09 projects.

These vessels fit to all standards of modern cruise and hotel industries. There are all types of necessary rest (restaurants, Spa centres, fitness, bars, etc.). All leading native cruise operators took part in development of vessels therefore concepts completely meet their requirements.

PV09, PV300 and PV300VD vessels are not only up-to-date high comfortable and handy vessels but also safe, ecology "clear" passenger carriers that fully meet requirements of all International (PV300VD vessel) and Russian requirements (PV09, PV300, PV300VD vessels).

PV09 river-sea cruise vessel "Shtandart" was laid down in 2014 and put into operation in 2017. The vessel is intended for cruise voyages through European inland waterways of Russia, including Volga-Balt, Volga-Don, Belomor-Baltiysk Canal, Volga river, Moscow River; also on lines between Moscow port, St.-Petersburg port and Belomorsk port through Belomor-Baltiysk Canal; with sailing to White Sea and to the region of Solovetsky Islands, Gulf of Finland; Caspian, Azov and Black Seas.

PV300 river-sea cruise vessel "Mustay Karim" will be operated on long cruise lines Moscow – St. Petersburg, Moscow – Astrakhan and Moscow – Rostov-on-Don with replacing classical cruise vessels of 301 and 302 projects. The vessel was laid down in 2017 and is planned to be built in the beginning of 2020.

PV300VD river-sea cruise vessel "Petr Velikiy" is intended for work on lines Moscow – St. Petersburg, Moscow – Rostov-on-Don – Sochi, Astrakhan – the ports of the Caspian Sea, and in Black and Mediterranean Seas. The vessel was laid down in 2016 and is planned to be built in 2021. PV300VD will be the only (as for 2019) river-sea vessel in Russia which meet requirements of International Conventions.

8. REFERENCES

1. EGOROV, A.G., 'Analysis of Decommissioning Trends of River Cruise Vessels and Prognosis of River Fleet Structure with Definition of Most Demanded Types of Vessels for Conversion and Modernization (in Russian)', *Journal "Morskoy Vestnik" 1(69), pp. 21-27*, 2019.
2. EGOROV, G.V., ILNITSKIY, I.A., KALUGIN, Ya.V., 'River-sea Cruise Passenger Vessels for Russian IWW (in Russian)', *Science-technical collected book of Russian Maritime Register of Shipping 36, pp. 255-276*, 2013.
3. EGOROV, G.V., ILNITSKIY, I.A., KALUGIN, Ya.V., 'Line-up of River Cruise Passenger Vessels (in Russian)', *Journal "Vestnik ONMU" 2(38), pp. 20-40*, 2013.
4. EGOROV, G.V., 'Analysis of European River Cruise Experience and Its Influence on Native Cruise Vessels of New Generation (in Russian)', *Journal "Morskoy Vestnik" 3(43), pp. 23-28*, 2012.

9. AUTHORS BIOGRAPHY

Dr Alexander Egorov holds the current position of chief researcher at Marine Engineering Bureau since 2009. He is responsible for all Bureau science researches, conference publications and for some non-typical calculations. His previous experience includes researches of existing river and river-sea passenger vessels, including operation experience.

Prof Gennadiy Egorov holds the current position of general director at Marine Engineering Bureau since 1995. He is responsible for organization of work of all Bureau. His previous experience includes researches of existing river and river-sea passenger vessels and creation of new concepts.

Igor Ilnitskiy holds the current position of 1st deputy director general and chief designer at Marine Engineering Bureau since 1998. He is responsible for projects of different types' vessels. His previous experience includes design of new concepts of passenger vessels and ferries.

INTRODUCTION OF NEW SHAPED SHIP HULL "Y-HULL"

Patent No: 10201901198Q[1]

Amitavo Chowdhury Wye, Managing Director, ShipDyn Ltd., Bangladesh

SUMMARY

Y-HULL is the hybrid of Mono-hull and Catamaran which generates reduced wave by the hull and results in low resistance while moving forward, compared to Mono-hull & Catamaran ships. Ship hulls generate significant waves on the river ways which create heavy pitching & rolling motion for small country boats which are used for crossing of rivers by people living in rural parts of third world countries. The invention of Y-HULL also reduces river bank erosion because of its unique shape combining Mono-hull and Catamaran. This hull experiences less resistance which increases efficiency and reduces fuel cost. This combined hull shape is called Y-HULL as the hull buttock-line has a shape of English letter **Y** at the centre keel position. This Y-Hull ship will significantly reduce accidents of small country boats caused by heavy waves induced by ship's hull. Y-Hull is patented from Intellectual Property Rights Office of Singapore (IPOS).

KEY WORDS
Ship Design, Ship Resistance, Shipbuilding

1. INTRODUCTION

The water ways of third world riverine countries are unsafe for several reasons. Usually river routes are used by commercial ships i.e. Cargo Ships, Oil Tankers, LPG Carriers and other different types of ships. But people living in the river areas use river routes to cross the river to save time by using small makeshift country boats made of wood with very low stability standard. Therefore, small boats plying on rivers encounter huge waves from big ships plying on the same route. Overloading of passengers is also a usual scenario in these parts of the world. The invention of Y-HULL relates to an approach of reducing induced waves generated by ship's hull which might be a solution to the accidents occurred due to induced wave by ship's hull.

Fig. 1.a: Typical example of passengers crossing river in coastal areas of Bangladesh

Fig. 1.a and Fig. 1.b illustrates the crossing of river by passengers in rural areas living in coastal area of Bangladesh. When a big ship sails through, huge water waves are generated which cause heavy rolling & pitching motion of these small boats. As these boats are usually over loaded, they capsize causing death of passengers. If these induced waves can be controlled or diminished, number of such accidents can dramatically be reduced. As a natural phenomenon these resulting waves cannot be diminished completely but a much more modernized hull shape can control/reduce induced waves.

Fig. 1.b: Typical example of passengers crossing river in coastal areas of Bangladesh

2. BACKGROUND OF THE RESEARCH

In a country like Bangladesh; low income people try to save as much money as they can. Therefore, they use small boats to cross rivers even if there is availability of bridge on the river. It saves money as well as time because of huge traffic congestion over bridges. As a result, small boats play a vital role to their livelihood. Around the clock, these people rely on small boats risking their lives for the sake of cost effective mode of transport.

Demand for small boats is huge and passenger capacity depends on the sizes. Usually one boat can accommodate 10-15 people. But practical data shows, these boats are always loaded with at least 25 passengers. The world around us has changed significantly over the years. Technology has become one of the main drivers of economic and social development. However, village areas of third world countries are yet to see the advancement. 100% of these boats are manually driven and sometimes driven by kids who have no experience and learnt from their forefather. Whenever a big wave is induced by commercial ships, these boats experience huge rolling and pitching motion. Commercial ships are designed for optimum speed and in the busy river routes they often move at their highest speed thus their hulls and propellers create larger waves. And these waves often result in massive accident.

A study shows in Bangladesh, almost 44% of accidents occur due to Adverse weather & Overloading and 9% due to excessive current. 54% of the affected ships are passenger ships and of these 10% are passenger travelers on small boats.

3. WAVE ENCOUNTER METHOD

Small boats encounter these waves at 90^0 which are even more dangerous. ***Fig. 2*** shows a typical diagram of direction of movement of Small boats and big ships and ***Fig. 3*** shows a diagram of Ship waves encountered by small boats.

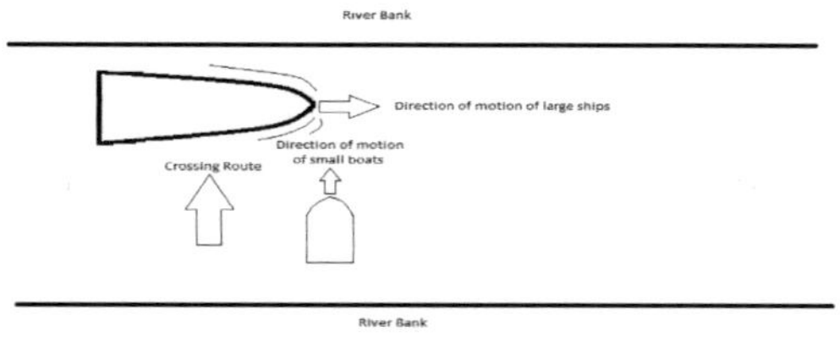

Fig. 2: Typical diagram showing direction of movement of Small boats and Commercial ships

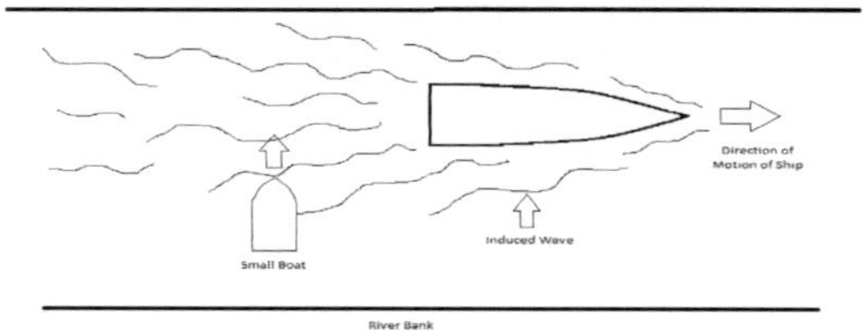

Fig. 3: Diagram of Ship waves encountered by small boats

4. DESIGN OF PROTOTYPES

4.1 PRINCIPAL PARTICULARS

Particular	Dimension		Unit
	Y-Hull	Mono-Hull	
Length Overall	13.50	15.20	Meters
Breadth	04.26	04.50	Meters
Depth	01.00	1.20	Meters
Draft	0.800	0.800	Meters
Superstructure height	03.20	3.20	Meters
Engine	110	110	Hp X 01
Propeller	34	34	Inch
Passenger Capacity	105	105	Person Seating

Table 1: Principal Particulars of Prototype

4.2 LINES PLAN OF Y-HULL

1. The ship has 27 stations in its lines plan where station 0 to 14 in Mono-hull which is the stern part and from station 15 to station 27 its catamaran which is bow portion of the ship, considering station 0 at the aft.
2. There is basically no connection point of Mono-hull and Catamaran rather the hull shape is gradually transformed in to Mono-hull from Catamaran from Bow to Stern. Therefore, it generate reduced wave.

4.3 DESIGN PROCESS

The design is finalized after studying several hull shapes of Mono-hull and Catamaran ships. Lines plans of these two hull shapes were thoroughly studied and an alternative hull shape was found which can reduce waves. Also, it was important to monitor the efficiency of the ship so that fuel consumption doesn't increase. A line plan combining Mono-hull and Catamaran was developed and studied. After analyzing the hull shape and lines plan, it was agreed that the Y-Hull has the capability to reduce the induced waves and also should not affect engine performance. After finalization of the lines plan, the hull was modeled, and stability of the hull was analyzed. A mono-hull with similar principal particulars was also designed, studied and analyzed.

4.4 ANALYSIS OF RESISTANCE & POWER

The target of the analysis was to check resistance of both the hulls. Resistance analysis in Maxsurf showed remarkable data.

	Mono-hull			Y-Hull		
Speed (Kt)	Froude Number (LWL)	Mono-hull Resistance (KN)	Mono-hull Power (kW)	Froude Number (LWL)	Combined Hull Resistance (KN)	Combined Hull Power (kW)
0	0	--	--	0	--	--
0.5	0.024	0	0.003	0.024	0	0.004
1	0.048	0	0.019	0.048	0.1	0.027
1.5	0.072	0.1	0.06	0.072	0.1	0.087
2	0.096	0.1	0.136	0.096	0.2	0.198
2.5	0.12	0.2	0.257	0.12	0.3	0.375
3	0.144	0.3	0.434	0.144	0.4	0.631
3.5	0.168	0.4	0.681	0.168	0.5	0.981
4	0.192	0.5	1.031	0.192	0.7	1.44
4.5	0.216	0.7	1.55	0.216	0.9	2.025
5	0.24	0.9	2.348	0.241	1.1	2.757
5.5	0.264	1.3	3.689	0.265	1.3	3.669
6	0.288	1.8	5.464	0.289	1.6	4.813
6.5	0.312	2.6	8.671	0.313	1.9	6.291
7	0.336	4.2	15.095	0.337	2.3	8.273
7.5	0.36	5.6	21.439	0.361	2.8	10.952
8	0.384	6.2	25.483	0.385	3.5	14.472
8.5	0.408	7.8	34.022	0.409	4.2	18.159
9	0.432	11.1	51.571	0.433	4.6	21.089
9.5	0.456	14.5	70.9	0.457	5	24.305
10	0.48	17.9	92.02	0.481	5.4	27.821
10.5	0.504	21.3	114.936	0.505	5.9	31.647
11	0.528	24.7	139.659	0.529	6.3	35.797
11.5	0.552	28.2	167.07	0.553	6.8	40.302
12	0.576	33.6	207.573	0.577	7.3	45.289
12.5	0.6	38.2	245.572	0.601	7.9	50.647
13	0.624	41.8	279.392	0.625	8.4	56.401
13.5	0.647	44.4	308.663	0.649	9	62.567
14	0.671	46.4	333.871	0.674	9.6	69.164
14.5	0.695	47.7	355.905	0.698	10.2	76.206
	Mono-hull			Y-Hull		
Speed (Kt)	Froude Number (LWL)	Mono-hull Resistance (KN)	Mono-hull Power (kW)	Froude Number (LWL)	Combined Hull Resistance (KN)	Combined Hull Power (kW)
15	0.719	48.7	375.728	0.722	10.8	83.708
15.5	0.743	49.4	394.214	0.746	11.5	91.681
16	0.767	50.1	412.078	0.77	12.2	100.139
16.5	0.791	50.6	429.872	0.794	12.9	109.093

Table 2: (Contd.) Comparison between Mono-hull and Y-hull in terms of Speed vs Resistance & Power

From the above data, it is observed that, resistance of Y-Hull is less than Mono-hull in each and every speed. The power required to drive Y-hull ship is also less compared to mono-hull ship of same particulars. These software generated data matches with practical data. At cruising speed of 9 Knots resistance of mono-hull is 41% higher than that of Y-Hull and required power is 40% higher for mono-hull. This indicates Y-Hull requires less amount of fuel to operate as well.

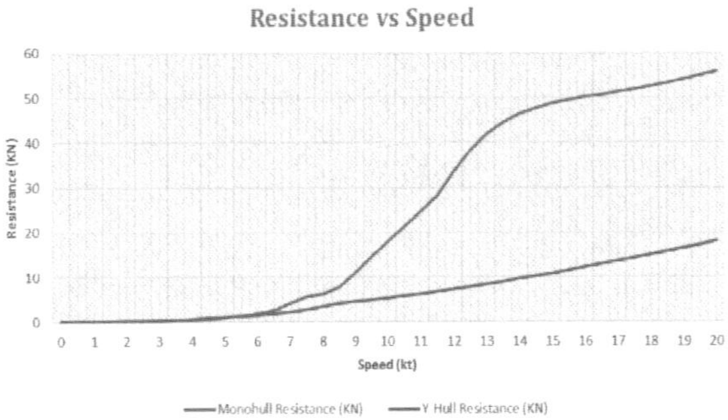

Graph 1: Graphical Comparison of Resistance between Mono-hull & Y-Hull

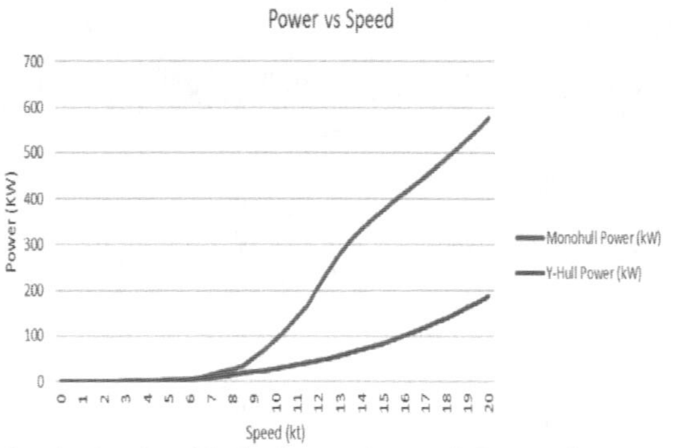

Graph 2: Graphical Comparison of Power between Mono-hull & Y-Hull

4.5 ANALYSIS OF WATERPLANE AREA OF Y-HULL

The result of the analysis of Water plane area is shown below with diagram:

Draft Amidships (m)	1.200	0.900	0.800	0.600	0.300	0.000
Water plane. Area m^2	25.360	25.040	23.896	21.689	13.452	4.562

Table 3; Results of Water plane area of Y-hull

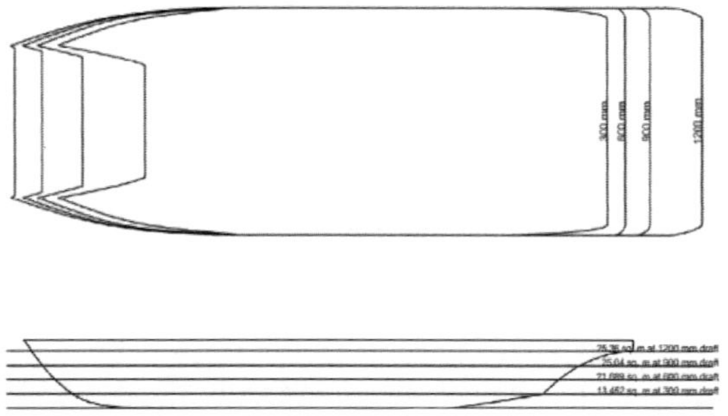

Fig. 4: Water plane area of Y-Hull

4.6 CONSTRUCTION

After finalization of all designs, construction was conducted in strict monitoring so that design cannot be altered during construction. After 23 days of round the clock construction work by 30 skilled workers and supervision by 02 Naval Architects, first prototype was launched into water. After 48 days from Day 1 of construction, the ship was tested, and trial was done.

4.7 PRACTICAL RESULT

Practical results matched with theoretical analysis. There was hardly any wave generating from the bow part of the hull at maximum speed. Propeller was the only source of wave generation of the prototype. In order to get accurate results, test and trial of the ship was done on confined water way at Hatirjheel situated in Dhaka, Bangladesh. Test & trial was done both in Light Ship and in fully loaded condition. And both cases showed tremendous result. The total passenger capacity is 105. Currently the boat is in commercial use for 08 months now.

4.8 MODIFICATION IN DESIGN

After successful test and trial of Prototype-1, design of the ship was checked again, and a new model was prepared and analyzed by adding a small form of bulbous bow keeping the principal particulars exactly same as prototype-1.

4.9 CONSTRUCTION OF PROTOTYPE-2 WITH MODIFIED DESIGN

After finalization of modified designs, construction of Prototype-2 was conducted. After 30 days of round the clock construction work by 23 skilled workers and supervision by 02 Naval Architects, prototype-2 was launched into water. After 62 days from Day 1 of construction, prototype-2 was tested and trial was done.

4.10 PRACTICAL RESULT OF PROTOTYPE-2

Practical results of Prototype-2 again matched with theoretical analysis. Reduced waves were induced compared to mono-hull from the bow part of the hull at maximum speed. Again, propeller was the only source of wave generation of the prototype. Test and trial of prototype-2 was also done on confined water way called Hatirjheel. Test & trial was done both in Light Ship and in Full load condition. The total passenger capacity is 105. Currently the boat is in commercial use for 04 months now.

4.11 MATERIALS AND METHODS

Shipbuilding MS Plate: Shipbuilding MS plate is used to build the ships.

Shipbuilding MS Angle: Shipbuilding MS Angle is used to build the ship's stiffeners and structural members.

Engine: 110HP ISUZU Diesel Engine is used to propel the ship.

Propeller: 01 No twin screw 03 Blade propeller of 34" diameter is used.

Gearbox: A 3:1 marine reduction gear is used as gear box.

Other materials: Standard shipbuilding materials were used to build this ship

5. OVERVIEW OF THE RESEARCH

After studying several hull shapes of several commercial vessels, this hull shape was designed. Series of experimentation and analysis were done before going into practical demonstration. The first boat of Y-Hull was launched into water on June 11, 2018 and has started commercial operation on July 01, 2018, after 48 days of round the clock construction process. After evaluating the performance of the ship, some modifications were introduced to the design. The new modified design was studied and analyzed and construction of second ship was started on July 20, 2018. The second ship was launched into water on August 20, 2018 after 30 days of round the clock construction process. The practical results of these two ships totally matched with expected results.

6. PHOTOGRAPHIC ILLUSTRATION

3D Design of Y-Hull

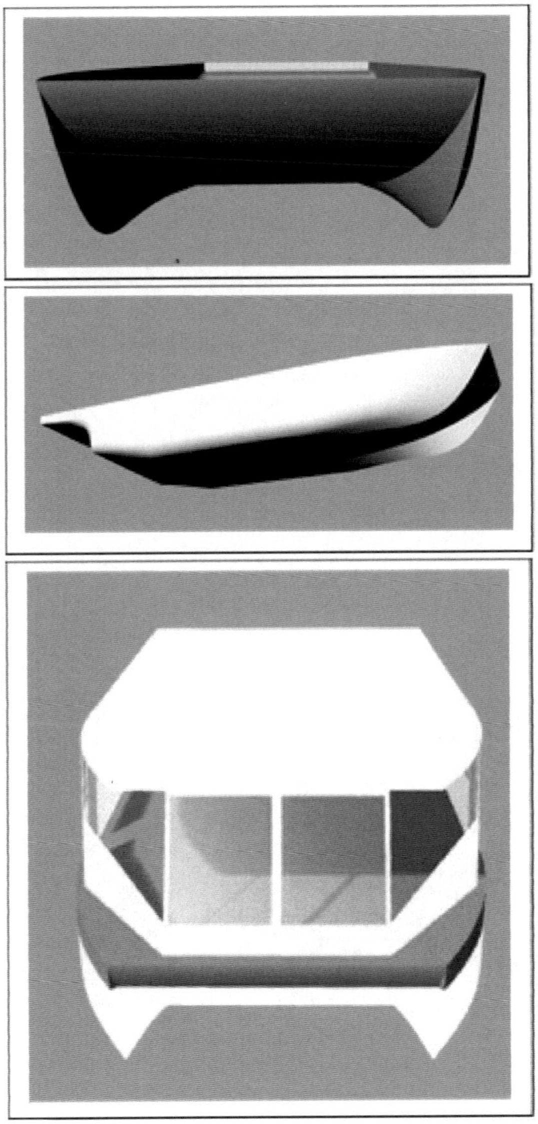

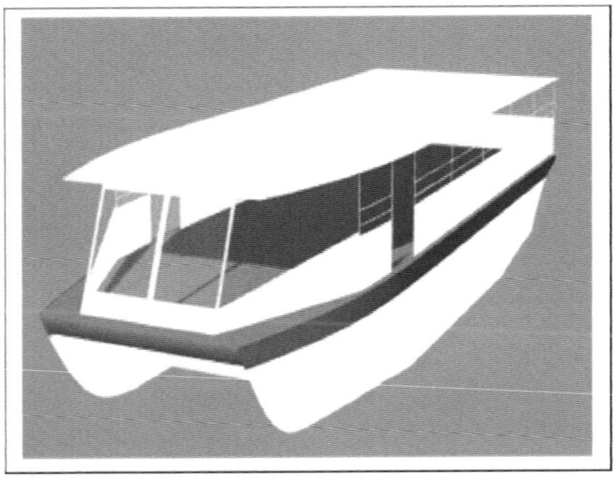

Fig. 5: 3D Model of Prototype 1

3D Design of Mono-hull

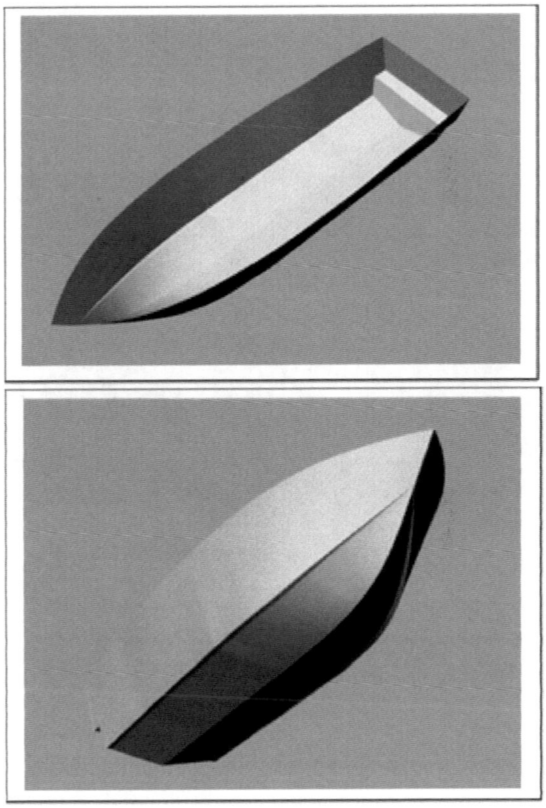

Fig. 6: 3D Model of Mono-hull Prototype

7. CONSTRUCTION PHOTOS

Fig. 6: Frame Construction of Prototype-1 (Photo Taken from bow)

Fig. 7: Shape of Hull of Prototype-1 (Photo Taken from bow)

Fig. 8: Frame Construction of Prototype-1 (Photo Taken from Aft)

Fig. 9: Trial at 50% Loaded Condition of Prototype-1 (No water wave generation from Bow)

Fig. 10: Port Side view of the Ship (Prototype-1) No water wave generation from Bow

Fig. 11: View of Mono-hull prototype

8. CONCLUSION

The invention of Y-Hull is described herein which is the new shaped hull hybrid of Mono-hull and Catamaran. It is proven that this hull generates reduced wave from the bow part without affecting the engine performance rather requiring less fuel. The hull is designed in such a way that the wave induced by the hull is reduced significantly which results in low resistance. In this paper, I tried to present the practical data of Y-Hull which is a new invention and patented under Intellectual Property Rights Office of Singapore. After studying hundreds of different ship hulls along with ten year experience in this field, I came up with Y-Hull which might be a solution of small country boat accidents which will surely reduce unwanted deaths. Y-hull will also significantly reduce river bank erosion which will save lots of money for the port authorities and Government of riverine countries.

9. ACKNOWLEDGMENT

The author gratefully acknowledges the workers who continuously worked during the construction of the Prototypes. The author also would like to thank two Naval Architects who worked under the author during construction phase of the prototypes.

10. ADDITIONAL INFORMATION

Opinions and calculations expressed here are solely author's experimented data. The entire results and findings are copyright project by Patenting under The Intellectual Property Rights Office of Singapore.

11. AUTHOR'S BIOGRAPHY

Amitavo Chowdhury Wye, holds current position of Managing Director at ShipDyn Ltd. He is responsible for entire company operation including Ship Design, Shipbuilding and Project Management of individual projects and leading a team of 08 Naval Architects, 02 Mechanical Engineers, and construction team of more than 50 people. He is experienced in ship design, shipbuilding and project management for 10 years. He designed the first catamaran ship designed and built in Bangladesh and he was also the Lead Designer of first LPG Carrier designed and built in Bangladesh.

AN EXPERIMENTAL INVESTIGATION INTO REDUCTION OF SLOSHING EFFECTS IN SWIMMING POOLS ONBOARD CRUISE VESSELS

D Nunes and **D Hudson**, Fluid Structure Interactions Group, Faculty of Engineering and Physical Sciences, University of Southampton, U.K.

SUMMARY

This study aims to analyse swimming pool sloshing for cruise vessels and suggest ways to mitigate its impacts without disrupting passenger usage. A custom-built large model-scale modular tank fitted to a 6-axis motion table was designed to conduct experiments. The motions experienced by the vessel in typical moderate were calculated and used as input to the 6-axis motion table. Tests were conducted to determine the natural frequencies and wave amplitude of the water in the pool for a typical modern large cruise vessel with and without baffles fitted. Baffle designs considered include designs based on those applied to LNG vessels and road tankers. These must be adapted for the different form factor of swimming pools and to satisfy drowning and entrapment regulations. As a result, large surface area baffles cannot be used practically.

NOMENCLATURE

b	Tank Width (m)
E	Motion Spectral Density (m^2 s rad^{-1})
g	Gravitational Acceleration (m s^{-2})
h	Filling Height (m)
$H_{1/3}$	Significant Wave Height (m)
L	Tank Length (m)
n	Mode of Vibration
σ	Standard Deviation
ρ	Density (kg/m^{-3})
RAO	Response Amplitude Operator
S	Wave Spectral Density (m^2 s rad^{-1})
T	Period (s)
T_1	Mean Wave Period (s)
R	Scaling Ratio
μ	Wave Heading (degrees)
ω	Angular Frequency (rad/s)

1. INTRODUCTION

1.1 OVERVIEW

Sloshing is defined as the motion of any fluid inside a container. This is caused when partially filled tanks are excited at certain frequencies [1]. Liquid sloshing motion can be used as a motion damper in some applications, such as passive motion control in tall buildings [2] and research vessels. Nonetheless, sloshing is particularly problematic in Liquefied Natural Gas (LNG) carriers where partially filled tanks can cause significant stability problems or load peaks on the tank sides [3].

On cruise ships, sloshing effects are an issue present in swimming pools [4]. Even though the liquid mass is generally small, compared to LNG tankers, the high position of the pool on the ship enlarges sloshing consequences. Free surface water motion on swimming pools onboard cruise ships can cause passenger discomfort, wet decks and ultimately stability issues [5]. If this was not a considerable problem until a few years ago, with the cruise industry's rapid growth a desire for safer operation and greater customer satisfaction rates has arisen. As a result, there is currently a significant interest in improving swimming pool designs.

1.2 AIMS AND OBJECTIVES

This paper aims to investigate potential non-intrusive methods of reducing sloshing in swimming pools onboard cruise ships through experimental testing. Moreover, it aims to create an experimental test method suitable for studying the impact of installing various features on the behaviour of water in a ship's swimming pool.

The objective is to develop a model scale testing tank which emulates current cruise ship pool designs. The model will be installed on a 6-axis motion shaker which will simulate the motions experienced by the pool when the ship moves in a seaway. The influence of depth changes and installing features along the pool will be assessed by evaluating water elevation, number of peaks and overall free-surface behaviour.

The paper reviews the implications of scaling the ship's pool to model scale. Baffle designs implemented on LNG tankers will be analysed together with their suitability for swimming pool usage. The experimental setup and its features are described, and the results of the model scale experiment presented.

2. METHODOLOGY

2.1 SLOSHING

Sloshing as a problem involves the prediction of pressure distribution, forces, moments and natural frequencies of the free surface within a container. The hydrodynamic pressure of liquid in a rigid container can be split into two different components, one proportional to the tank's acceleration, influenced by the moving liquid velocity being the same as that of the tank. The other, a "convective" pressure, which represents the free surface motion [1]. Sloshing has an infinite number of natural frequencies; however, the lowest modes are those generally excited by the ship's motion [6]. The sloshing behaviour is generally classified according to filling ratio, type of impact and fluid behaviour [7]. There are four main impact types (Figure 1).

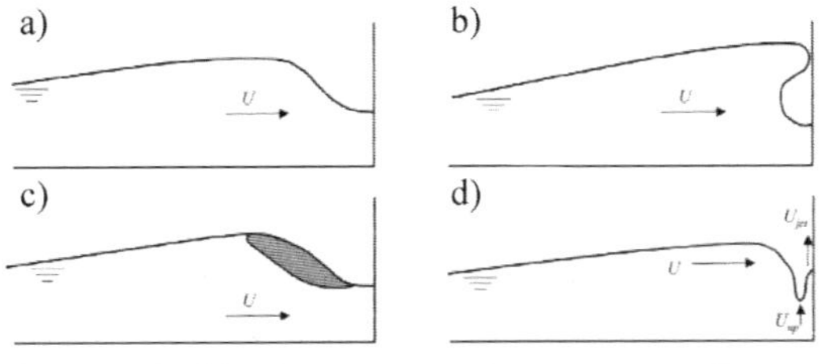

Figure 1: Sloshing impact types: a) steep waves, b) breaking waves, c) aerated waves and d) flip-through [7].

Potential flow analysis may be used to predict the sloshing response [7] but it has difficulties doing so when great dissipation is involved [6]. The dynamic transverse force due to liquid sloshing induced by translatory motion of a rectangular tank is [7]:

$$\frac{F}{\rho g h L b} = \ddot{x}(t)\,\frac{1}{g}\left[1 + \sum_{n=0}^{\infty} \frac{8 \tanh\left\{(2n+1)\pi\frac{h}{L}\right\}}{\pi^3 (2n+1)^3 \frac{h}{L}}\,\frac{1}{\frac{\omega_n{}^2}{\omega}-1}\right] \quad (1)$$

where $\ddot{x}(t)$ is the tank acceleration. From equation (1) it is possible to observe that the closer the excitation frequency is to the resonance frequency of the tank, the greater is the system's response.

On pools onboard cruise ships, unlike in LNG tankers [6], higher motions are normally observed under pitch and surge motions [4]. This is due to the orientation of the pools. At larger fill depths, as seen in cruise ship pools, large standing waves are observed near the resonance frequency [1].

Methods used to analyse sloshing include experimental testing, computational fluid dynamics (CFD) [8] and theoretical methods. For this study experimental methods are employed.

2.2 SCALING

The main problem that arises with scaling a tank from full scale to model scale is that not all parameters of interest (Table 1) scale similarly. For instance, sloshing pressure is difficult to scale due to conflicting scaling laws [9].

Table 1: Parameters that affect sloshing

P	Pressure on tank
g	Gravity
L	Tank length
h	Filling Height
b	Tank width
x_0	Excitation amplitude
T	Excitation period
μ	Viscosity
ρ	Density
σ	Surface tension
c	Speed of sound in water
E	Elastic modulus of tank
ΔP	Pressure difference (free surface – vapour)

Sloshing experiments are typically carried at a constant Froude number. Hence, the pool's characteristic length and speed are considered. This will enable for a more accurate visual comparison between model and full scale.

The relationship between inertia and viscous forces, given by the Reynolds number, is of importance, as flow viscous forces play an important part in the flow damping around baffles, but taken as secondary to the free surface shape. The latter will be key to define improvements in flow behaviour. Additionally, it may be of interest to understand the effect of surface tension especially in the beach area. If the Bond number is greater than unity, it can be neglected when modelling the free surface [7].

2.3 MOTION DATA GENERATION

2.3 (a) Ship Response

In this paper, the response to wave excitation is considered since this is most likely to be the dominant cause of ship motions.

In order to assess the behaviour of the pool when the ship is travelling in a seaway, an understanding of the ship response needs to be developed. To generate a response spectrum at the centre of gravity, the wave spectra for a particular sea state is combined with the ship's response amplitude operator (*RAO*) for a particular motion (2) (Figure 2).

$$E_i(\omega) = RAO_i^2(\omega)S(\omega) \quad (2)$$

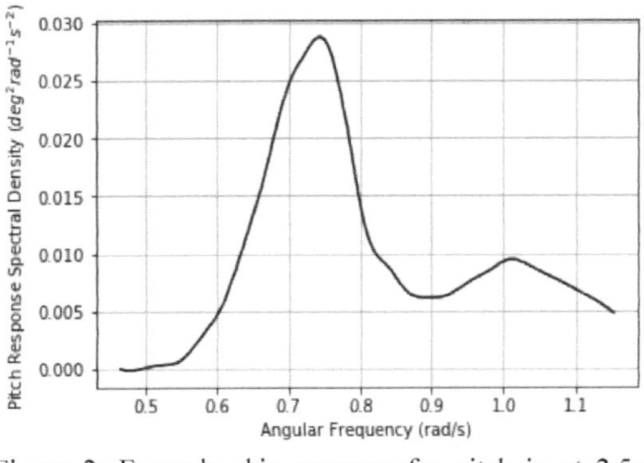

Figure 2: Example ship response for pitch in at 2.5m significant wave height and 8.5 mean zero crossing period.

The vessel's operational speed was taken in this study to be 20.8 knots and both head (μ=180°) and bow-quartering seas (μ=135°) were considered.

In this study the two-parameter, Bretschneider spectrum is adopted (3). Significant wave height and zero crossing period are taken from Global Wave Statistics by British Maritime Technology [10].

$$S(\omega) = \frac{A}{\omega^5}\exp\left(-\frac{B}{\omega^4}\right) \quad (3)$$

$$\text{with } A = 0.25H_{1/3}^2 \text{ and } B = \frac{691}{T_1^4}$$

In order that the sloshing analysis is performed for a range of scenarios both frequent and more extreme sea states on the Atlantic Ocean, a typical cruise vessel route, are considered, as summarised in table 2. The more extreme sea states are likely to cause sloshing.

Table 2: Sea States adopted in model scale testing.

	Significant Wave Height (m)	Significant Wave Period (s)	Heading
Sea State 1	2.5	8.5	180°
Sea State 2	3.5	10.5	180°
Sea State 3	2.5	8.5	135°

The ship's RAOs were provided by the ship owner, as derived from model scale experiments prior to construction.

2.3 (b) Frequency to Time Domain

The input for the 6-axis shaker platform is a time series of the motion amplitude, for each degree of freedom separately. Since the motion analysis is performed in the frequency domain, the generated response spectrum generated is translated to the time domain [11].

The signal is then scaled from ship to model scale and resampled to generate a smooth signal and hence motion from the 6-axis shaker. A sample signal is shown in Figure 3.

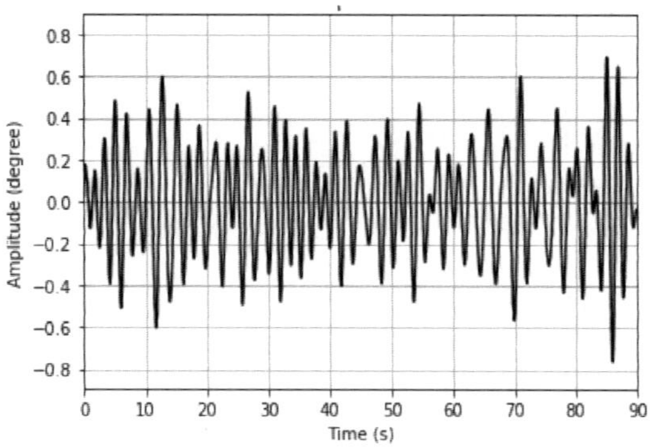

Figure 3: Signal generated for pitch motion for Seastate1.

The seakeeping and motions analysis are performed for the centre of gravity of the ship. The motions at any location of interest may be derived from those at the centre of gravity.

To obtain the motions at the pool location trigonometry can be used to transform the surge, heave and pitch motions into the vertical and horizontal motion.

2.4 POOL DESIGN

From analysis of pool dimensions on a range of ships it is observed that they are generally rectangular with a ratio of length to width in the order of 2:1 and the depth varying from 1.5 to 2.3 metres. Additionally, beach areas around the pool are becoming increasingly popular and larger. These areas can have a water depth up to 5 cm and aim to create the impression of larger pool areas as well as increasing the area where people can "contact" the water. Taking this into account, and after data collection for a modern cruise ship, the basis pool parameters were defined (Table 3).

Table 3: Pool Parameters.

Length	11 m
Width	4.9 m
Height	1.4 - 2.25 m
Beach Area$_s$	72 m^2
Vertical distance to LCF	44 m
Horizontal distance to LCF	21 m

It is important to note that swimming pool design on cruise ships must consider the vessel's response in surge and pitch since these are the motions most likely to excite the pool's water. If the frequency of response matches the natural frequency of the water motion in the pool, undesirable high amplitude waves may be observed. The nth natural frequency for a full scale rectangular-shaped two-dimensional tank can be predicted from (4) [13].

$$T_{N,n} = \frac{2\pi}{\sqrt{\frac{gn\pi}{l}\tanh\left(\frac{n\pi h}{l}\right)}} \quad (4)$$

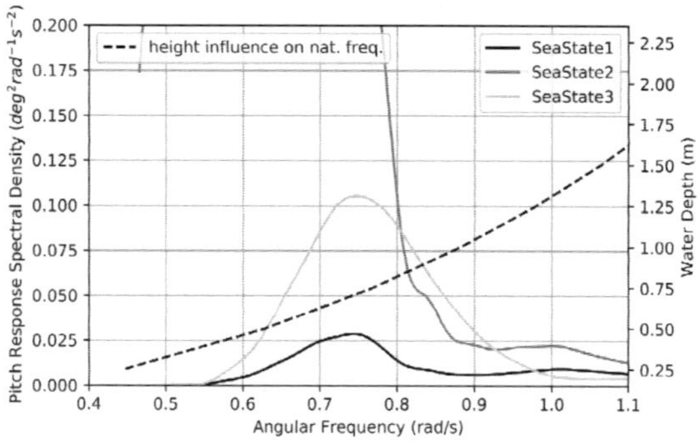

Figure 4: Comparison between ship response for the 3 seastates studied and natural frequency of the pool for a range of heights as predicted by (4).

For the ship considered the response spectra is compared, in figure 4, to the natural frequency of the pool, predicted for different water depths. From figure 4, it can be seen that as the depth decreases, so does the natural frequency. The choice of pool depth as greater than 1.5m may be seen to reduce the likelihood of the natural frequency matching the ship's natural frequency.

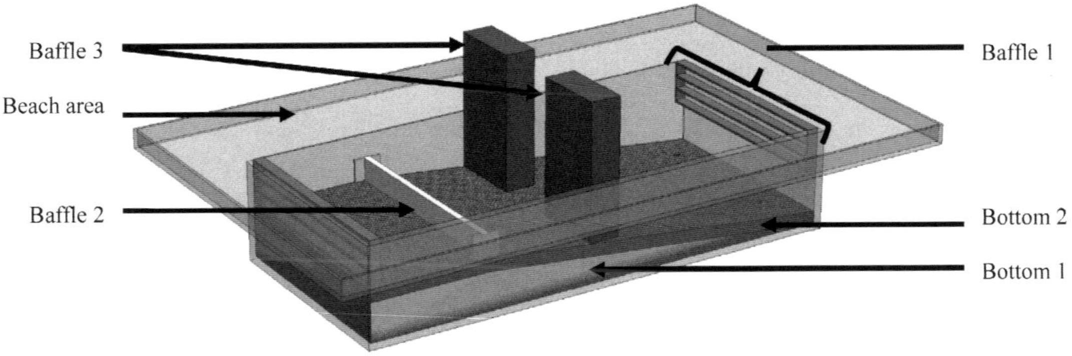

Figure 5: CAD design of the basis pool with the bottoms and baffles installed

2.5 SLOSHING REDUCTION METHODS

For cruise ships, sloshing mainly affects the pool's operation and safety, rather than ship stability or structural integrity. Currently, pool depths tend to be larger than desirable for a leisure pool in order to detune the pool's natural frequency from the ship's response (Figure 4). With the frequencies detuned, the pool could be operational in most sea states, however, this depth is quite large for passenger safety and comfort. At pool depths of 2.25m, according to the NCD Risk Factor Collaboration network height distribution [14], most humans cannot reach the pool's bottom. At a minimum depth of 1.4m most children would likewise struggle to reach the bottom. There is therefore a desire to reduce water depths without compromising pool operability. In order to realise this, sloshing suppression devices could be applied to current pool designs and their effectiveness tested at shallower pool depths.

2.5 (c) Suppression devices examined

There are issues raised with downscaling LNG baffle design and detuning pool natural frequencies by changing overall dimensions. Taking this into account, the three baffle designs examined in this study attempt to incorporate the principles adopted in LNG tanks into feasible designs for a swimming pool environment. This means that these have to be safe according to entrapment and drowning regulations [15], as non-intrusive for users as possible and not limit the pool's operability at any condition.

The first suppression device, referred to as *baffle 1,* was designed aiming to compensate the width limitation of baffles by introducing a larger array of them close to the free surface (Figure 5). The closer to the free-surface the horizontal baffles are placed the more effective these are [6]. In practise, as the water moves along the pool's wall, the vortices generated at each baffle's tip, due to viscosity, should dissipate energy, hence reducing sloshing amplitude. Additionally, under higher flow velocities they would work additionally as an air trapping mechanism [2]. These work by generating air pocket regions which create damping due to density differences, adding to the absorption of kinetic energy due to the baffle array. Previous studies indicate that with ideal ratios sloshing reductions can be in the order of 50% [17]. The device was designed with the following specification, referring to K. Hyunjong [17], l/L% = 3%, where l is the protrusion width of the baffle. The gap between baffles was chosen to be the minimum permitted by entrapment regulations [16].

The second sloshing suppression device proposed, hereby referred to as *baffle 2,* consists of an underwater vertical barrier (Figure 5) which is likely to change the natural frequency of the fluid by, in certain instances reducing, the pool's effective length as well as generating turbulence at the barriers tip if close enough to the free-surface [18]. At H/h >0.2 there is hydrodynamic damping in the form of blockage effects on liquid convection, where H is the baffle's height. These effects increase with the ratio, whereas the vortex at the baffle's tip loses its predominance until the baffle's width is higher than the free-surface and no vortex is observed. For this reason, the ratio H/h% was chosen to be 70%.

The third suppression device, referred to as *baffle 3,* attempts to realise the benefits of physically dividing the pool whilst tackling its main disadvantage, the reduction of the pool's effective length. To do so, two barriers are installed at either side of the pool's midpoint, leaving enough width for people to swim between each pool end (Figure 5). It is important to bear in mind that it is likely that the changes in the natural frequency will not be as pronounced as with a complete physical barrier [4]. The barriers have a full-scale width of 1.7m and length of 0.7m.

2.6 EXPERIMENTAL TESTING SET-UP

The principal aim of the model scale testing is to measure sloshing natural frequency and analyse the free surface characteristics for the various pool and baffle arrangements. It is simultaneously desirable to create a testing setup which may be a reliable basis for future systematic testing of the impact of pool features for sloshing, or other performance parameters.

The model scale was chosen based on operational constraints. The bigger the scale, the closer dynamic similarity is obtained with the actual ship and the lower the testing frequencies are, which is likely to be more practical for the testing rig. However, the building complexity, required materials and budget increase with a larger model. The scale was taken as 1:12.5 (Table 4). Perspex [19] was selected as the building material as it provides sufficiently rigid boundaries with the possibility to observe the water movement.

Table 4: Model Scale Pool Parameters.

Length	0.880 m
Width	0.392 m
Height	0.172 m
Beach Area$_s$	0.46 m^2

One key aspect of the model scale pool is that its basis design has a flat bottom at 0.18m (corresponding to 2.25m full scale). It is very unlikely that a pool deeper than this would be installed on a ship. Therefore, any bottom configuration may be attached, allowing for different bottom shapes to be studied. For this experiment, a bottom emulating existing pools was fabricated, as well as a bottom which allows for a shallower pool with the same basic geometric profile.

To comply with operational safety guidelines for the testing platform, a water retention tank was built to contain any spillages from the model pool (Figure 6).

2.6 (a) Testing Platform

In order to accurately reproduce the motions experienced by the swimming pool, a 6-axis motion table was required as a testing platform. The Human Factors Research Unit at the University of Southampton has a large 6-axis motion table and this was selected, with the characteristics shown in Table 5. The motion table can perform a range of motions from simple harmonic motion in one axis to irregular motion simultaneously in all axes. This testing platform ensures that the motions experienced at any given location on a ship can be reproduced accurately.

Table 5: 6-axis motion table characteristics

Displacement	1m vertical
	0.5m fore and aft
	0.5 lateral
	20-degree roll, pitch, yaw
Acceleration	+- 10ms^{-2} translational
	+- 5 rads^{-2} rotational
Frequency range	0-50Hz
Platform dimensions	2m x 3m
Payload	1 tonne

2.6 (b) Testing Procedure

Initially, the natural frequencies of the model pool at different filling ratios were assessed by exciting the tank in surge at a range of frequencies close to the expected natural frequency, as predicted by equation (4). An amplitude of 0.1m was applied for these tests. The resonant frequency is determined as the one with the greatest internal, wave amplitude. These resonant frequencies are compared to the prediction. This procedure is repeated each time a new feature is installed in the model pool, either a baffle or bottom. The free-surface behaviour for each of three different sea states is analysed for each featured installed. Each run took 8m30s which correspond to 30 minutes in full-scale. The arrangements tested are illustrated in Table 6.

Video is recorded for all runs at two distinct locations. Water elevation in the model pool is measured at three locations (centre and each end point) using 330mm twin wire resistance wave probes. Data for wave elevation is recorded at 100Hz. Figure 6 shows a photograph of the experimental arrangement of the model pool on the motion table.

In order to video record the experiment two GoPro Hero4 Black digital cameras were used. One was set to record at 2K (30 fps) resolution, with a wide-angle field of view and located so that it could capture the whole tank. The second was set to record at 1080p (60fps) resolution, with narrow field of view and positioned to capture local flow around baffles or the pool walls.

Figure 6: Experimental Set-Up.

Table 6: Tested arrangements. Waterline marked with black line.

Arrang.	Diagram
Flat Shallow	
Flat	
Flat Deep	
Bottom1	
Bottom2	
Baffle1	
Baffle2	
Baffle3	

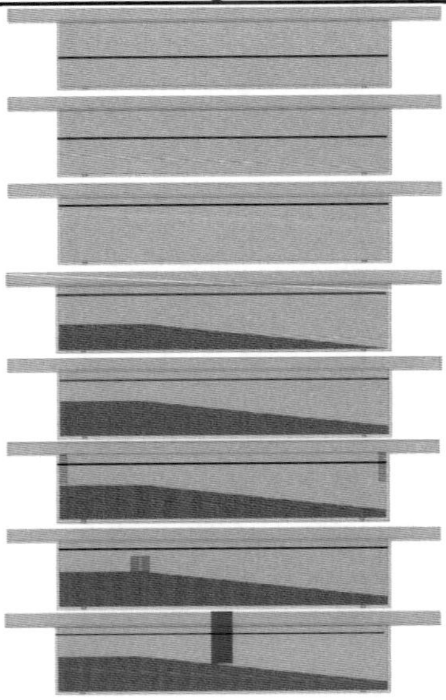

3. RESULTS AND DISCUSSION

Comparisons between the theoretical and experimental values of natural frequency for a rectangular shaped tank with different water depths are made in this section. The impact of the suppression devices on the natural frequency and free surface behaviour of the pool are also assessed, for both regular and irregular sea-states and with different pool water depths.

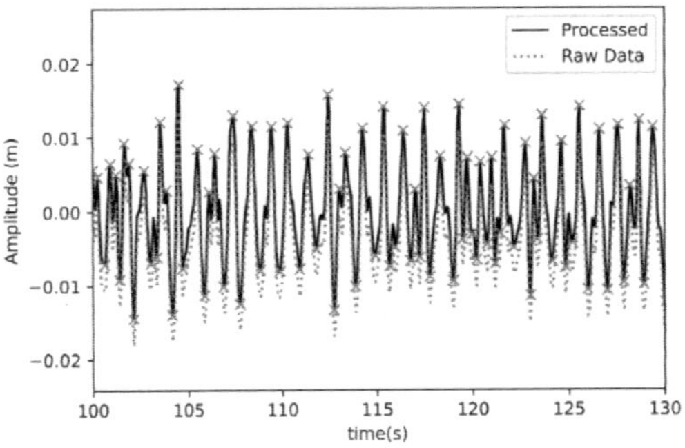

Figure 7: Signal processing for run with current bottom and no baffle installed at seastate 1 for the middle wave probe.

The signal acquired by the wave probes was detrended linearly, in order that water spillages do not affect the results. This is achieved by subtracting the result of a linear least-squares fit to the data from the original (raw) data. Recorded peak wave amplitudes were filtered so as to remove any peak with a height smaller than 10% of the maximum peak height. This ensures that small ripples, which would be unlikely to have a significant impact on pool users, are not included in the analysis. Figure 7 presents results for this signal processing for a run with the current pool bottom and no baffles, in seastate 1.

3.1 NATURAL FREQUENCY

3.1 (a) Experimental versus Theoretical

The natural frequency of the system was taken as the frequency for which the average maximum amplitude across the three wave probes was the highest. Results are show in Table 7 and Figure 8 and show good agreement (within 2%) between the prediction for the shallow and middle filling depth cases for a flat pool bottom. However, once the tank is filled to the top (Flat Deep) and the waves are able to dissipate to the beach area, the natural frequency is 7% lower than that predicted using equation (4). This difference is likely to be caused by the theoretical value assuming a rectangular shaped tank, where the water is confined by the vertical wall sides.

Table 7: Experimental Natural Frequencies compared to theoretical predictions.

Arrang.	Depth (m)	Exp. N. F (Hz)	Theor. N. F. (Hz)	% dif.
Flat Shallow	0.08	0.5	0.4976	0.48
Flat	0.115	0.6	0.5874	2.12
Flat Deep	0.14	0.595	0.6400	7.29
Bottom1	0.128	0.5525	0.6160	10.86
Bottom2	0.106	0.51	0.5662	10.44

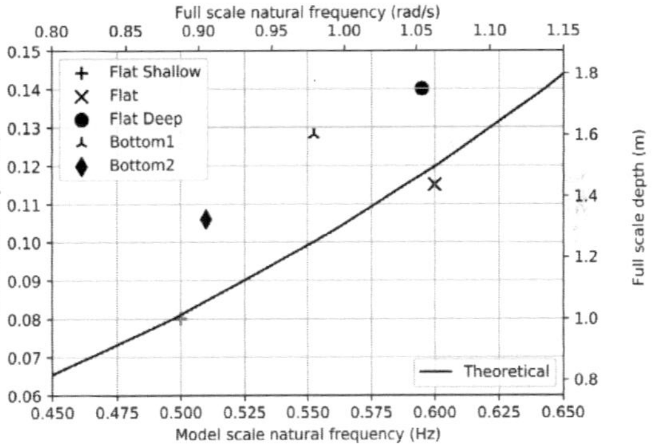

Figure 8: Experimental natural frequency for each tank configuration and varying depth versus the theoretical prediction (4). The depth on the inclined bottom is taken as the mean depth along the length.

3.1 (b) Suppression Devices

Comparing the response of the pool water for the different installed suppression devices it is possible to categorise the results into amplitude reduction and changes in natural frequency.

Figure 9 presents the measured model scale wave amplitude for the different baffle designs and frequencies tested. Comparing to the condition with no baffle, Baffle 1 resulted in an 8% reduction in the measured wave amplitude (Figure 9). The reasons behind this minor reduction are discussed in detail in section 3.2(b). The device also appears to have no influence on the natural frequency of the water motion. Similarly, Baffle 2 also does not appear to alter the natural frequency significantly but does produce a much bigger reduction in measured wave amplitude, at around 50% (Figure 9). Baffle 3 shows the largest reductions in wave amplitude, at around 88%. There is also no natural frequency detected within the range of frequencies tested (Figure 9). According to the theoretical prediction, the natural frequency for this configuration would be higher since the pool is practically divided into two and hence has a much smaller effective length for the purposes of the natural frequency.

The performance of these baffles when the pool is undergoing pure surge motions serves as an indicator of their performance in an irregular sea state. It is, however, important to note that pitch will have an influence on the flow behaviour and there will also be effects caused by the randomness of the motion in a realistic seaway.

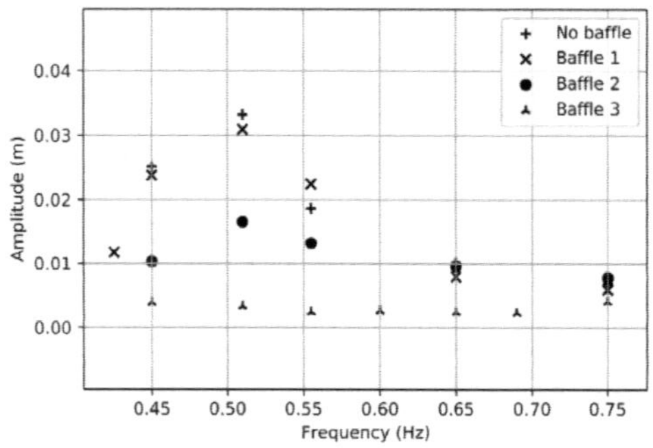

Figure 9: Effect of suppression devices used on model scale wave amplitudes versus the range of tested frequencies with Bottom 2 in place.

3.2 BEHAVIOUR IN A SEASTATE

3.2 (a) Overview

Table 8: Sum of peak numbers and % difference to Bottom 2 arrangement across the wave probes.

Arrang.	**Peak Number**	**% dif. to Bottom2**
Bottom1	3006	-6
Bottom2	3194	---
Baffle1	2612	-18
Baffle2	2532	-21
Baffle3	2143	-33

Table 9: Sum of average peak amplitude and maximum peak amplitude and % difference to Bottom 2 arrangement across the wave probes. Average standard deviation of the peak amplitudes across the wave probes.

Arrang.	**Mean peak amp. (mm)**	**% dif. to B2**	**Max. peak amp. (mm)**	**% dif. to B2**	**σ (mm)**	**% dif. to B2**
Bottom1	44	0.9	140	-9.7	8.04	-3
Bottom2	44	---	155	-	8.32	-
Baffle1	41	-7	122	-21	7.81	-6
Baffle2	19	-57	53	-66	3.19	-62
Baffle3	13	-70	39	-80	2.50	-70

The behaviour of the different pool arrangements in seastate 1 will be assessed for the number of peak wave amplitudes recorded, the average amplitudes of the highest 90% of peak wave amplitudes, the maximum peak wave amplitude and the average standard deviation of the peak wave amplitudes.

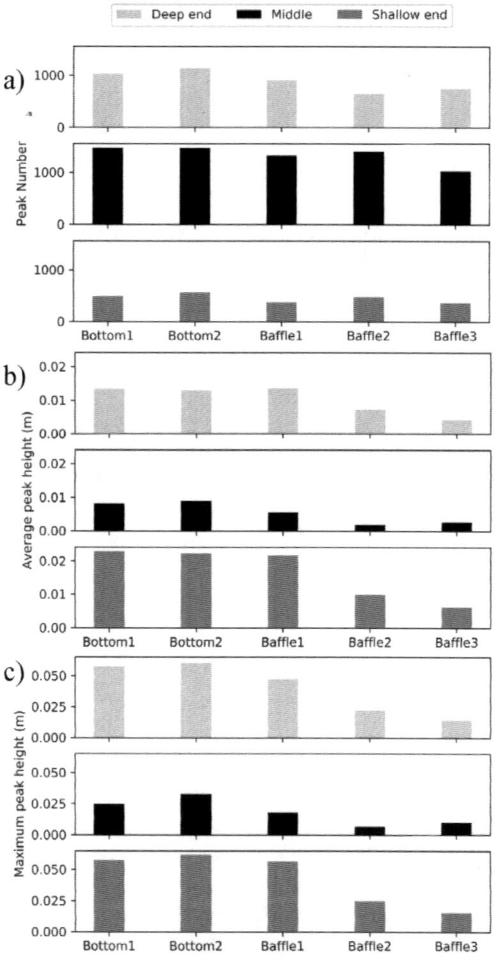

Figure 10: a) Peak Number; b) Average peak amplitude and c) maximum peak amplitude for each probe and arrangement.

Overall, with bottom 2 installed there are 6% more peaks (Table 8) and the maximum peak wave amplitude is greater compared to the current bottom depth in place. On the contrary, the average amplitude of the wave peaks is effectively the same for this condition, compared to bottom 1, as is the standard deviation (Table 9). For this reason, it can be concluded that for seastate 1, the decrease in pool water depth does not cause a significant negative impact on the overall free surface behaviour. This agrees with the natural frequency predictions and ship's response to Seastate 1 drawn in Figure 4.

In general, all baffles showed reductions in the parameters evaluated. It should be noted that baffle 3 shows the most pronounced improvements.

Figure 10 shows the total number of peaks recorded, the average peak wave amplitude and the maximum peak wave amplitude recorded for each wave probe and baffle arrangement. The higher average peak wave amplitude at the shallow end (Figure 10b) may be explained by the decrease in depth in this region. On the other hand, the decrease in number of peaks is related to the longer wave period due to the change in pool's natural frequency (Figure 10a).

Regarding sloshing impact modes, under Seastate 1, only steep waves were observed. However, it is important to note that under harsher seastates aerated and breaking waves were occasionally observed.

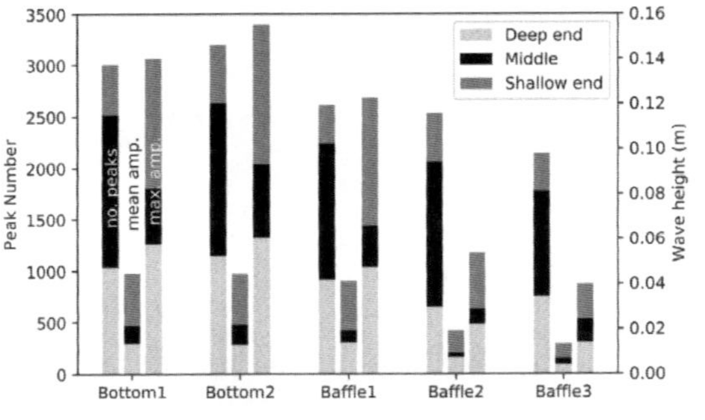

Figure 11: Combined results of peak number, average peak height and maximum peak height for Sea State 1

3.2 (b) Baffle 1

Despite a small reduction in amplitude being observed when finding the surge natural frequency (Figure 9), overall Baffle 1 shows a more significant reduction in the number of peaks, an average peak height reduction at the middle of the pool and reduction in the maximum peak at all wave probes (Table 8 and 9). This shows that the effects of raising the bottom could be mitigated by the implementation of this baffle, but the water motion in the pool would still be severe in the analysed sea state.

The small variation seen in the average peak wave amplitude recorded by the shallow and deep end probes is due to the flow behaviour close to the baffle (Figure 10b). Figure 12 shows that the flow, instead of splashing to the beach, is pushed up around the location of the wave probe, hence higher amplitudes are measured.

Baffle 1 shows desired improvements in the free surface behaviour as splashes and spillages are less likely. As discussed, this suppression device works by dissipating energy close to the side wall. This reduces the wave amplitudes, as seen by the average and maximum amplitude reductions at the middle wave probe (Figure 10b/c). Nonetheless, there is little influence on the coupled pitch-surge natural frequency of the pool. For this reason, the same basic free surface shape is maintained, as shown in figure 13a/b.

Regarding passenger safety and usability of the pool with this baffle installed, these potentially pose a risk to swimmers if a person is placed between an incoming wave and the pool wall, then they may be hurt from hitting the baffle's protrusions.

Since the performance of this baffle is not outstanding, and safety concerns may arise from implementing such a device, it does not seem that this device should be implemented in pools onboard cruise ships.

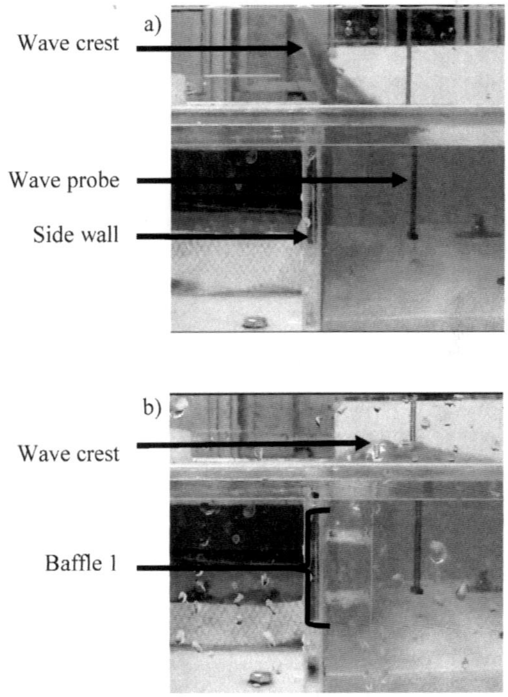

Figure 12: Water behaviour near the pool's side wall without any baffle (top) and with baffle1 (bottom) at an interval of 0.1s.

3.2 (c) Baffle 2

The installation of this device results in an overall reduction of all measured parameters to a greater extent

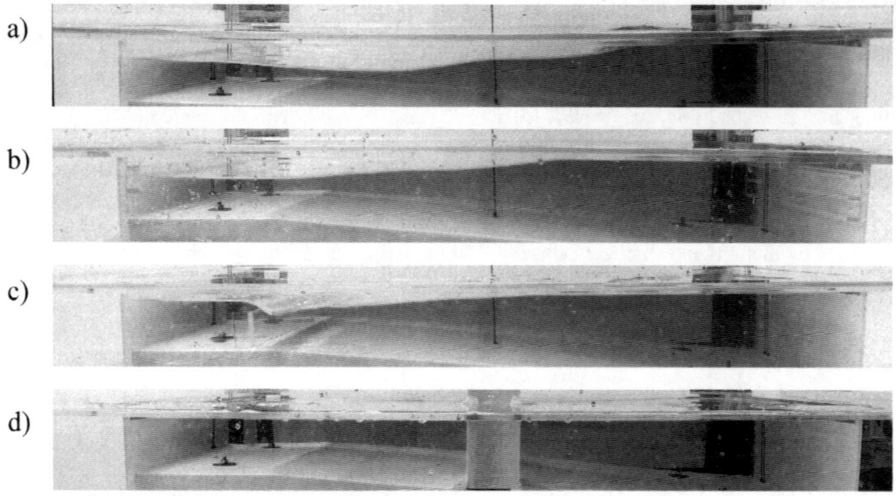

Figure 13: Free surface profile at the same instant under Sea State 1. a) Bottom2; b) Baffle 1; c) Baffle 2 and d) Baffle 3

than seen with baffle 1. The implementation of such a baffle, in a shallower pool, would improve the free-surface behaviour. In general, barely no splashes on the side walls nor spillages were observed in testing.

Looking at wave probe measurements individually, it is important to note that the middle wave probe registered very little average vertical displacement compared to the remaining probes (Figure 10b). This means that the number of traveling waves is greatly reduced compared to the other arrangements and standing waves with their node at the middle of the pool are more frequent. This was corroborated by visual observations. This suggests that for this seastate the natural frequency of the pool is excited by the ship's response at certain instances. However, high amplitude waves are not generated as there is energy dissipation by the vortices generated at the tip of the baffle. Even though very little change in the natural frequency was observed with this baffle for pure surge motion, with coupled surge and pitch motions the baffle appears to influence the natural frequency of the water in the pool.

In contrast to baffle 1, this arrangement not only reduced the average and maximum peak wave heights (Table 9), but also changed the free surface shape along the pool as can be seen in Figure 13a/c.

The working principles of this suppression device are simultaneously the reason for its success and its operational and safety drawback. The vortices generated at its tip dissipate energy to a great extent, as may be observed in Figure 14. However, these vortices can create significant downward forces, potentially causing users to be pulled toward the pool's bottom. This was not evaluated further in this study, hence additional testing, full-scale computational fluid dynamics and/or full-scale mock-ups would need to be performed to evaluate these effects more fully. Additionally, for larger motions, a wave travelling from the deep end breaks near the baffle and again on the pool wall side. For this reason, the effectiveness of this device appears compromised in harsh seastates.

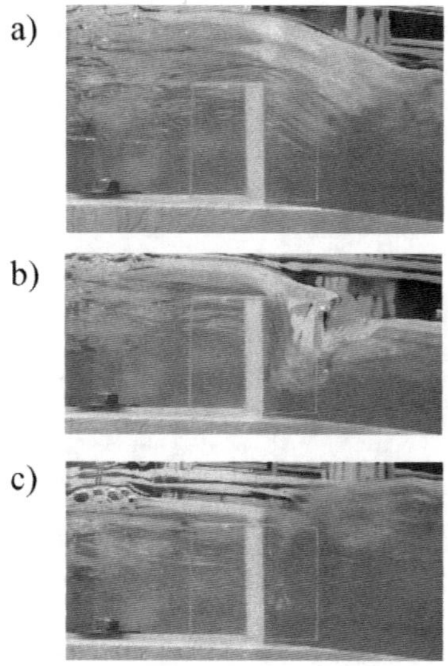

Figure 14: Water behaviour near the baffle region at intervals of 0.3s. On a) flow is towards the deep end; on b) the vortex sheet is fully developed; on c) flow is towards the shallow end.

This baffle could potentially be effective for certain seastates, if properly dimensioned and if further investigation is performed into the health and safety impacts. Furthermore, this baffle is not ideal in terms of minimising passenger disruption. For it to be effective it must be relatively close to the free surface, hence jeopardizing the free movement of users across the full length of the pool.

3.2 (d) Baffle 3

From the results presented in Tables 8 and 9, Baffle 3 appears to perform the best for all parameters measured. Additionally, the standard deviation of the peak wave amplitudes is least for this arrangement. This indicates that a smooth surface profile is achieved. From visual observation there are were no waves generated but only small ripples.

When looking at results for each wave probe it is observed that of all arrangements this is the one with the least difference in recorded data across the probes. This shows that water moves smoothly across the pool, with no chaotic behaviour near the side walls. One interesting observation is that a destructive interference between the wave systems within the pool is created . This interference adds significant damping to the flow (Figure 13d).

Unlike baffles 1 and 2, baffle 3 appears to be equally effective in both harsher and quartering seas. However, as previously seen, further investigations would have to be made in order to fully understand the strength of the vortices generated in harsher seastates. One possible way of reducing these would be to round the inside edges of the baffle. As this baffle does not work on the principle of energy dissipation through vortices, but through a large change of the natural frequency, reducing the vortices strength should not impact the effectiveness of the device.

In terms of safety and operability, this baffle design seems to be effective under most seastates and does not significantly constrain the usability of the pool. This baffle also enables the pool waterline to be raised into the beach level without significantly increasing sloshing risks. It is thought that the baffle could easily be incorporated as an aesthetic feature of the pool.

4. RESULTS' RELIABILITY

Repeat measurements were made for some arrangements in order to increase the degree of confidence in the results.

Figure 15 presents the wave amplitude time history for two instances of the same sea state. This demonstrates that there is close agreement between the measurements. Not only do the peaks and troughs of the wave amplitude match but also, the amplitudes are similar throughout. Figure 16 illustrates repeat data for several frequencies for

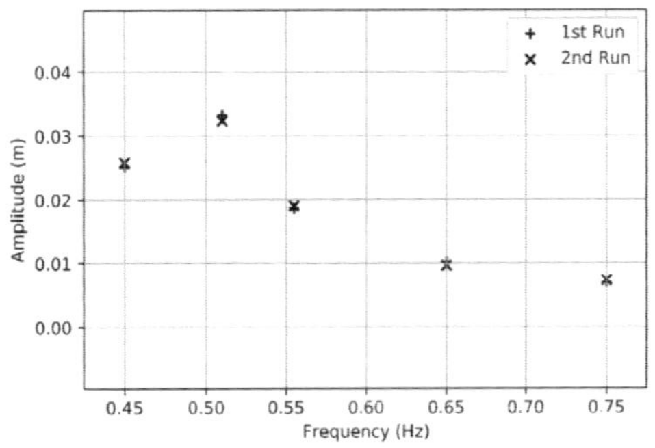

Figure 16: Natural Frequency plot for two distinct runs at two different day of bottom 2 arrangement.

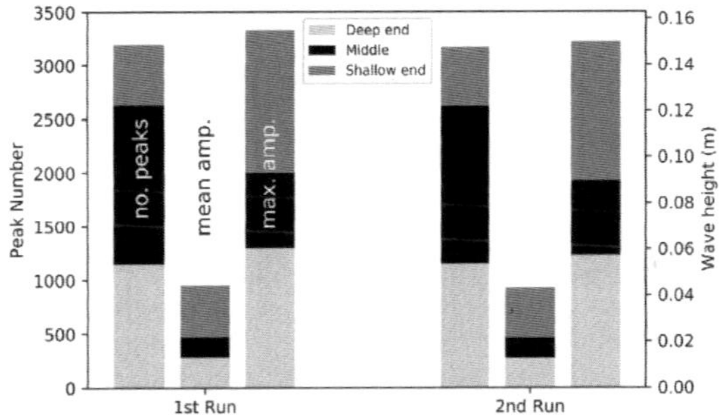

Figure 17: Combined results of peak number, average peak height and maximum peak height for two distinct runs at two different day of bottom 2 arrangement under Sea State 1.

the bottom 2 arrangement. This also shows close agreement between the two runs, within 3.5% difference. Results for two instances of the irregular sea state 1 are shown in Figure 17. The number of wave amplitude peaks, average wave amplitude peak and maximum wave amplitude recorded are very similar between the two runs.

It should be noted that the results between these two instances may have been affected by minor parallax errors when setting the water depth and so may not be identical. Additionally, wave probe measurements are dependent on the water temperature, as this changes the conductivity of the water. Platform displacements and accelerations were monitored and were within 1%-2% of the input signal for all cases.

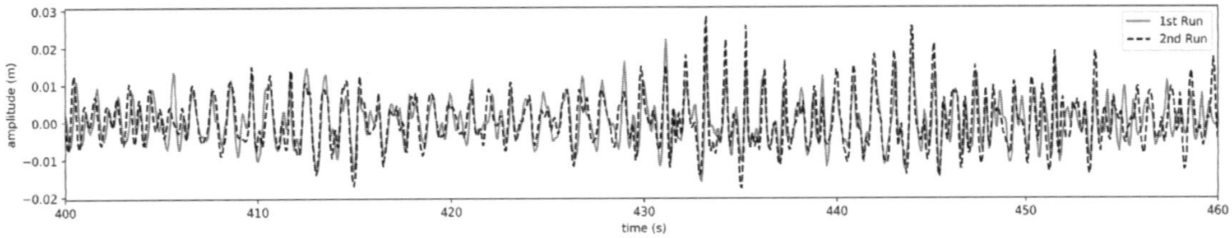

Figure 15: Comparison of two distinct runs at two different day of bottom 1 arrangement.

Overall, there is a good agreement between repetitions and the potential remaining errors should have very little impact in the overall results. The experimental testing was undertaken at full scale Froude number and hence viscous and surface tension effects are not representative of full scale.

5. CONCLUSIONS

A model-scale swimming pool experiment was developed to determine natural frequencies of pool water motion. The model-scale swimming pool is fixed to a large 6-axis motion platform in order that measurements of water motion in the pool with it undergoing representative vessel motions may be made. A flat bottom pool is tested, as is a bottom typical of those seen on modern cruise vessels. Three baffle arrangements were designed considering pool safety regulations and tested in order to judge their effectiveness in a range of sea-states.

From these tests it is seen that for a rectangular tank with vertical side walls that span the range of water motion, the natural frequency may be predicted accurately using a theoretical formula based on inviscid wave theory. If a beach area or an angled bottom are present, then the natural frequency is decreased for the same mean water depth. A correction factor could potentially be developed in other to provide a prediction for these cases.

The influence of a decreased water depth is dependent on the sea state and ship heading. Nonetheless, in general a decreased depth results in increased pool sloshing response. One way to mitigate the impacts of decreasing the water depth is to implement passive suppression devices, or baffles, in the pool. These should not affect the usability of the pool, nor compromise users in terms of drowning and entrapment.

Passive suppression devices either change the natural frequency of the system and hence change the response of the pool water to a given sea-state, or dissipate the energy associated with the pool water motion. Baffles that change the natural frequency of the water in the pool are seen to perform better. Energy dissipation devices have greater local impacts on the flow in the pool, but little impact on the overall behaviour in general.

A partial separation of the pool into two halves is seen to decrease the average, maximum and standard deviation of measured wave amplitudes. The number of peak wave amplitudes measured is also decreased in the sea-state evaluated. Further investigations should be performed on the vortices generated in the vicinity of this baffle. The applicability of these findings to different sea states, different pool locations and ship speeds and heading angles should also be studied.

6. ACKNOWLEDGEMENTS

This research would not have been possible without the support and data provided by the vessel operator. The authors would like to express their gratitude to the personnel of the University of Southampton Human Factors Research Unit for their help and support in running the experimental testing.

7. REFERENCES

1. IBRAHIM, R.A., Liquid Sloshing Dynamics: Theory and Applications, 2005.

2. CAVALAGLI, N., BISCARINI, C., FACCI, A., UBERTINI, S., and UBERTINI, F., 'Experimental and numerical analysis of energy dissipation in a sloshing absorber', *Journal of Fluids and Structures*, 2017

3. BROSSET, L., MRAVAK, Z., KAMINSKI, M., COLLINS, S., and FINNIGAN, T., 'Overview of Sloshel project', 2009

4. RUPONEN, P., MATUSLAK, J., LUUKKONEN, J., and HUS, M., 'Experimental study on the behavior of a swimming pool onboard a large passenger ship', *Marine Technology and SNAME News*, 2009

5. MONTERO, F., 'Sloshing in Swimming Pools', 2015

6. LU, L., JIANG, S., ZHAO, M., AND TANG, G., 'Two-dimensional viscous numerical simulation of liquid sloshing in rectangular tank with/without baffles and comparison with potential flow solutions', *Ocean Engineering*, 2015

7. GODDERIDGE, B., 'A phenomenological rapid sloshing model for use as an operator guidance system on Liquified Natural Gas carriers', PhD, University of Southampton, 2009

8. LLOYD'S REGISTER, 'Comparative sloshing analysis of LNG ship containment systems', *ShipRight Additional Design Procedures*, 2005

9. ABRAMSON, H., 'The dynamic behavior of liquids in moving containers, with applications to space vehicle technology', *Technical Report SP-106*, 1966

10. HOGBEN, N., DACUNHA, N. M. C., OLLIVER, G. F., ' Global Wave Statistics', 1986

11. LLOYD, A. R. J. M., 'Seakeeping - ship behaviour in rough weather', 1998

12. CELEBI, M., and AKYILDIZ, H., 'Nonlinear modelling of liquid sloshing in a moving rectangular tank', *Ocean Engineering*, 2002

13. MONTFORT, J. H., 'Sloshing the challenge of swimming pools on yachts', 2015

14. NCD RISC, 'Height – Evolution of adult height over time', *http://ncdrisc.org/height-mean-distribution.html*

15. SCARSI, G., 'Natural frequencies of viscous liquids in rectangular tanks', 1971

16. 'Swimming Pool Equipment', *BSI Standards Publication,* 2011

17. HYUNJONG, K., NANJUNDAN, P., YOON-HWAN, C., and YEON-WON, L., 'Reduction of sloshing effects in a rectangular tank through an air-trapping mechanism – a numerical study', Journal *of Mechanical Science and Technology*, 2017

18. HAKAN, A., 'A numerical study of the effects of the vertical baffle on liquid sloshing in 2D rectangular tank', *Journal of Sound and Vibration*, 2011

19. PANIGRAHY, P.K., SAHA, U.K., and MAITY, D., 'Experimental studies on sloshing behavior due to horizontal movement of liquids in baffled tanks', *Ocean Engineering*, 2009

9. AUTHORS BIOGRAPHY

Diogo Nunes is a 3rd year undergraduate student in Ship Science with Naval Architecture in the Faculty of Engineering and Physical Sciences at The University of Southampton, U.K. The work presented formed the topic of his Part 3 individual research project.

Dominic Hudson is the Shell Professor of Ship Safety and Efficiency at the University of Southampton. He has research interests in all aspects of ship hydrodynamics, performance management and operational efficiency. He has published over 110 papers in journals and proceedings of refereed conferences. He is a Chartered Engineer and member of RINA. He is presently a member of the 29th ITTC Specialist Committee on Ships in Operation at Sea and the RINA IMO Committee.

SPEED: A NEW SOFTWARE FOR SPEED & POWER SEA TRIAL ANALYSIS

P Becchi, CETENA S.p.A., Italy

SUMMARY

The last international recommended procedures and normative prescribe how the speed & power sea trials should be carried out and analyzed in order to evaluate the effective ship performances at the contractual load condition and keeping out the environment effect, that means wave, wind, current and water temperature. For this reason, CETENA recently developed the new software "SPEED" aimed to perform the analysis of the data measured during official S&P sea trials and to provide the final results just onboard as the sea trials have been finalized. In order to guarantee both the software development and the accuracy of computation that can be carried out, in summer 2018 the software certification has been completed in accordance with the rules ABS and RINA.
The ship speed and power performance analysed and thus referred to ideal condition can be used for verifying the Energy Efficiency Design Index (EEDI) of the ship.

NOMENCLATURE

L_{BP}	Length between perpendiculars (m)
Δ	Displacement (t)

1. INTRODUCTION

At the moment of a ship delivery, the need to establish the relationship between the ship speed, shaft power and propeller rate of revolutions under contractual condition requires well defined procedure to be followed, being the evaluation of ship performances strongly affected by some not well defined factors. Furthermore, the use of an accurate methodology leads to both the check of the fulfilment of the contractual obligations between builder and owner and the evaluation of the real ship performances, that represents a really important reference data set from the design point of view.
Procedures for the analysis of sea trials data have recently been published by international institutes/organizations like ISO [1] and ITTC [2-4] in order to give standard methodologies for the execution for speed/power trials, the analysis of the collected data and finally the definition of the expected ship performances at the contractual load condition, that usually is different from the trial one. Furthermore, the possibility to analyze the data and print an official report showing the final results just onboard at the conclusion of the tests represents a good feedback for both the owner and the builder.

2. CETENA' S SEA TRIAL SYSTEM

With over 40 years of experience in sea trials and experimental activities, CETENA has the combination of knowledge, skills and operational experience on board which contributes to give the clients the best solution in each phase of the ship's life. From the point of view of ship performance, in the last two years CETENA updated and implemented its own system for conduction, acquisition and analysis of the speed & power sea trials. The whole system consists of two independent modules that are interfaced each other for the data transfer:

- SYSMAN II that is the acquisition system and includes both hardware and software elements
- SPEED that is the analysis tool designed for the S&P analysis in accordance with the last international prescriptions

As mentioned above, the two modules are completely independent of each other in order to give the user the possibility to not be forced in using both of the two. In some cases, it could be asked to analyse the data with a different method or it could be also possible to perform the analysis of data acquired through a different acquisition system. In any case, there is the possibility to transfer all measured run data from SYSMAN II to SPEED speeding up the results analysis and avoiding error in the manual data entry.
Figure 1 shows the main console of the SYSMAN system, regarding the data acquisition during sea trial execution.
All software modules have been developed in LabVIEW® 2015, stating the large experience of CETENA in developing onboard hardware and software systems through the National Instruments development framework. The hardware of the acquisition unit has been selected and assembled by CETENA itself, saving weight and dimensions as possible. On the other side, the functionalities related to acquisition, connectivity and installation have been improved with respect to the previous version.
The system is able to acquire and store all kinds of signal (analog and digital) required for the analysis of both speed and power performances and manoeuvrability characteristics.
The recorded data are the results of a sampling procedure along 1 second, the sample rate of each signal depends on the type of the signal itself.
Together with the data required for the analysis, it is also possible to acquire any kind of analog signal (both current and voltage) that could be useful to know such as fuel consumption, propeller pitch, electric power and so on. In addition, the system is able to acquire N signals on serial lines (ModBus, NMEA, etc) together with environment conditions obtained by wave radar (if available) and anemometer that are really important for the propulsive performance analysis.

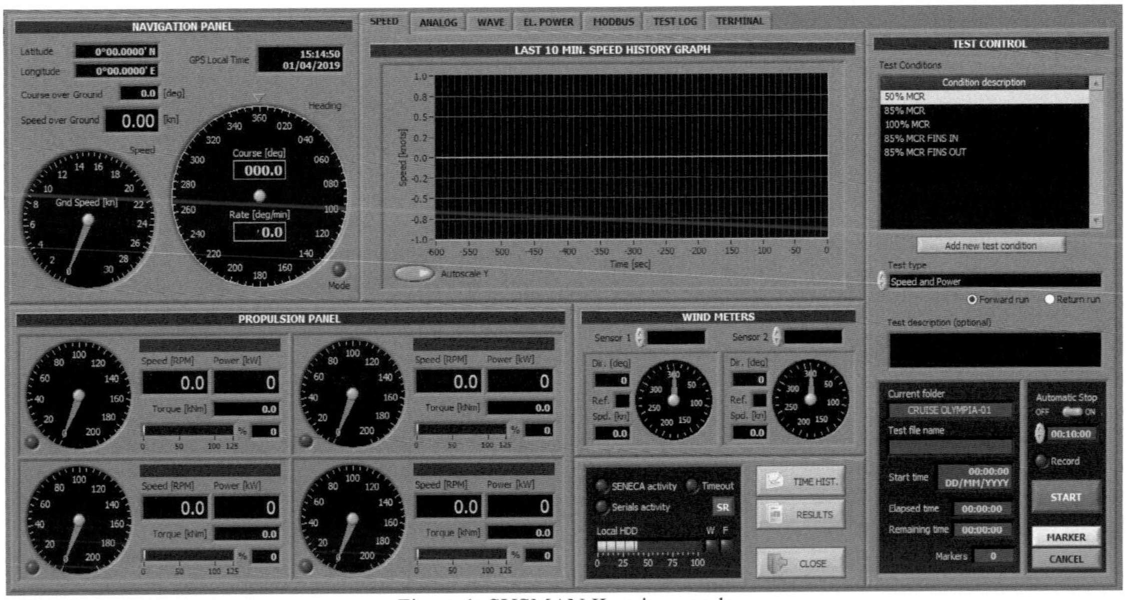

Figure 1: SYSMAN II main console

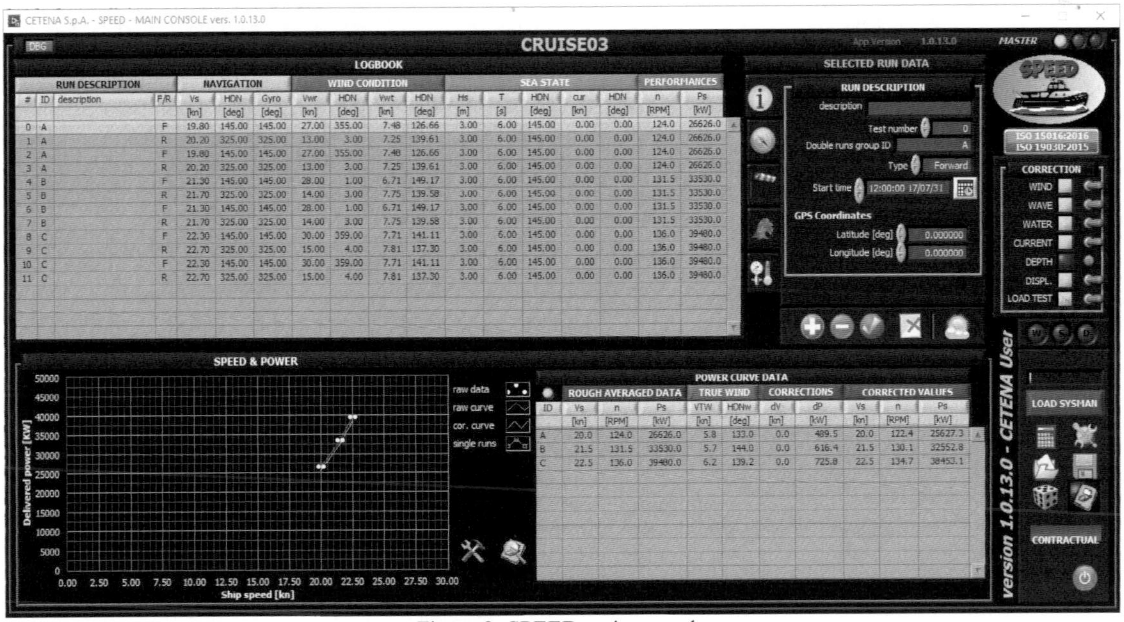

Figure 2: SPEED main console

In accordance with the referred procedures ([1]-[4]), the Speed&Power tests are carried out acquiring a minimum set of parameters for each run, that is: time, propeller shaft torque, propeller shaft speed, ship's position and heading, ship's speed over ground, relative wind condition and, if possible, sea state. All the parameter needed are acquired by the same hardware in the same time, in order to guarantee data simultaneity. Each single signal is managed depending on the sensors used, and one measured values data set is provided every second.

In August 2018 the development of the software SPEED (Figure 2) has been finalized together with its certifications. As previously mentioned, this software is aimed to carry out the S&P sea trials analysis, the computation of the speed-power curve in contractual condition and finally the printing of the technical report. The development of the software has been approached in order to provide to the user the possibility to choose one of the known international algorithms for the S&P analysis, that is: ITTC 2014, ITTC 2017, ISO 19030 and

ISO 15016. In this way it is possible to perform computation for a ship using the same algorithm adopted in the past, like in case of a twin ship. Another important aspect of the software is that the algorithm development and formulation provided CETENA the possibility to accurately study the normative prescribed and discuss about them with Italian Naval Register (RINA) during the validation and classification phase.

3. SPEED MAIN CAPABILITIES

The software SPEED has been developed keeping into account the following main required capabilities:

- easiness of the user interface
- easiness and reduced time requirement for the ship configuration
- use of spreadsheet models (Excel®) for the configuration file and model test data
- direct interface with the acquisition system
- possibility to choose the normative to be used
- possibility to choose the corrections to be activated
- possibility to check weather condition directly through the Virtual Weather Station (VWS) developed by CETENA for the storage and the query of environmental condition acting on sea/oceans
- reduced computation time
- results report in pdf format
- certification from the point of view of both design approach and results accuracy

These skills help the user, and especially the design team, to check the expected ship performances before the sea trials and speeding up the software configuration passing through Microsoft Excel® models as usually used for any design activity. On the other side, it is also possible for the owner representative onboard to control anytime the configuration used, including the model test data (Figure 4), the correction enabled and the measured run results. Furthermore, the main console always shows information about the numerical approach configured by the user, that is:

- the international normative adopted for multiple measured runs (ITTC 2014, ITTC 2017, ISO 15016) and also for single runs (ITTC 19030)
- the correction enabled for the analysis, and depending on the configuration completeness
- runs filters activation status

All these information are also included in the final report, in order to describe the conditions under which the measured runs have been filtered, corrected and analyzed. Figure 3 represents the information shown on the main console: the blue button is the control to be used for the selection of the procedure to be adopted, the green led and controls are related to the available corrections and their activation status while the last three yellow indicator show the status of the three main filters: wind, sea state and depth.

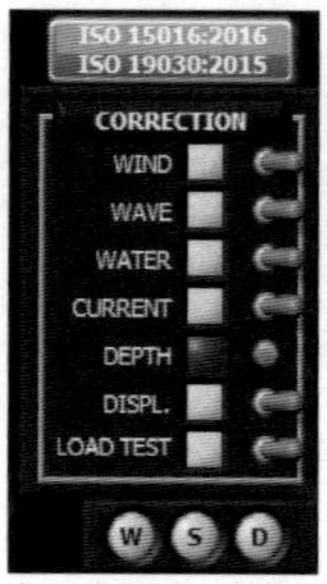

Figure 3: SPEED main console, information controls

The normative prescribes also that the two reciprocal runs of a single power setting to be parallel each other within one ship's length of the trial baseline. The combined use of SYSMAN II and SPEED makes it possible to check also the fulfilment of this requirement being also the GPS track recorded during the measurement.

The software has a custom interface with the acquisition system "SYSMAN II" that directly loads all the data acquired and shows the average data for each measured run performed. This skill guarantees no human error due to transcription from the acquired data system and the analysis one.

As prescribed by the international recommended procedures, the configuration data set includes the names of the three persons attending the sea trial representative team (owner, shipyard and sea trial measurement company) that will ask to sign the analysis report just after the execution of the speed and power tests, in order to provide to all the subjects involved in the sea trials an official report including both corrected measured data and the effective contractual ship performances.

Once the analysis has been finalized, the S&P curve in contractual condition can be checked through a custom user console (Figure 5) and the user has the possibility to query the whole ship performance (speed, power, rate of revolutions) simply through the cursor dropping along the curve shown.

Being designed and developed by CETENA itself, this software is continuously checked, updated/upgraded and implemented depending on clients or CETENA own requests.

4. INTERNATIONAL PROCEDURE IMPLEMENTED

As previously mentioned, the analysis of collected data is carried out consistently with the current international recommended procedures and regulations. Furthermore, the normative concerning ship in usual operability [5] has also been included in order to perform comparison with the others implying traditional multiple measured runs. It has been decided to not update the first ITTC computation module, but to leave it available (2014) and implement the new one (2017) in order to give the possibility to perform analysis also through the previous procedure release. Sometimes, it could be required to carry out the S&P analysis with an old version being the same procedure adopted for a twin ship some years before. In this way it is hence possible to have consistent results for twin ships and to check the differences with the new one procedure.

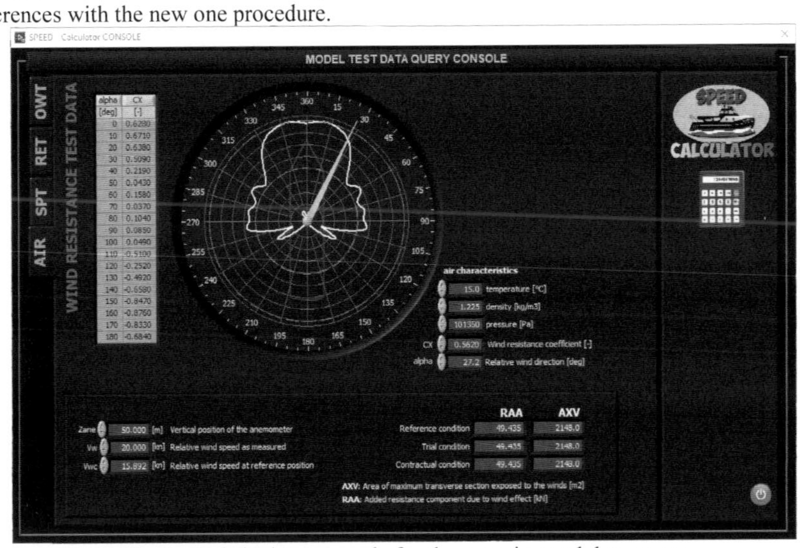

Figure 4: Calculator console for the experimental data query

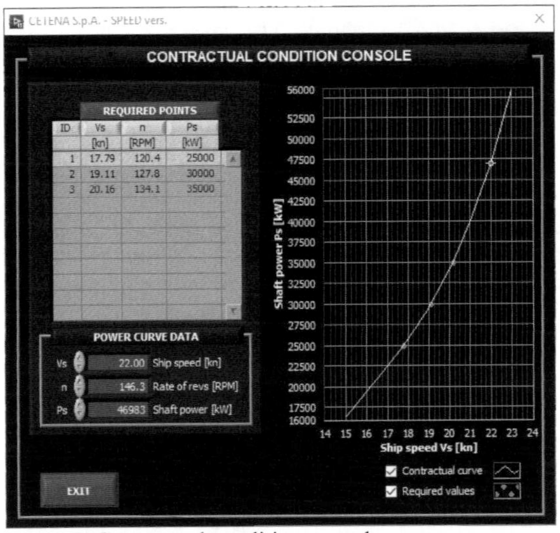

Figure 5: Contractual condition console

Furthermore, it has been decided to give the possibility to switch on/off every single correction prescribed. This skill is just for design point of view, providing a good feedback about the effect of each correction on the ship performances.

At the moment, the analysis methods implemented are the following:

- ITTC 2014
- ITTC 2017
- ISO 15016
- ISO 19030

The normative adopted for the correction modules implementation shows some differences especially related to the computation sequence, some formulations and/or variables to be used. While ISO 15016 and ITTC 2017 are very close each other, the ITTC 2014 looks to be different from some points of views, like the added resistance due to wind effect, the sequence of some corrections inside the overall flowchart and the formulation of the directional wave spectrum.

CETENA will keep the computations libraries updated with any future modification of these regulation and, if necessary, will improve new regulations if any. Figure 6 shows the analysis flowchart on which the ISO 15016 is based on, including the rule of the model test data. It can be seen the analysis starts with the definition of arrangement and sorting of the multiple runs, to be carried out at constant power setting. The analysis proceeds with the evaluation of the three main added resistance components due to wind, waves and water density. For the computation of all theses added resistances, it is possible to use experimental data obtained from custom model scale tests, that are: resistance test, self propulsion tests, seakeeping test and wind resistance test. In case of lack of the last two, it is possible to adopt different computation approaches, as suggested by the reference normative:

- wind resistance
  - o Regression formula by Fujiwara et al.
  - o Data set on the wind resistance coefficient
- wave resistance
  - o Direct correction method STAWAVE-1
  - o Empirical transfer function STAWAVE-2

After that the ship speed is corrected for both current and deep water effect and then the power in ideal condition is computed passing through the correction for propeller load and displacement.

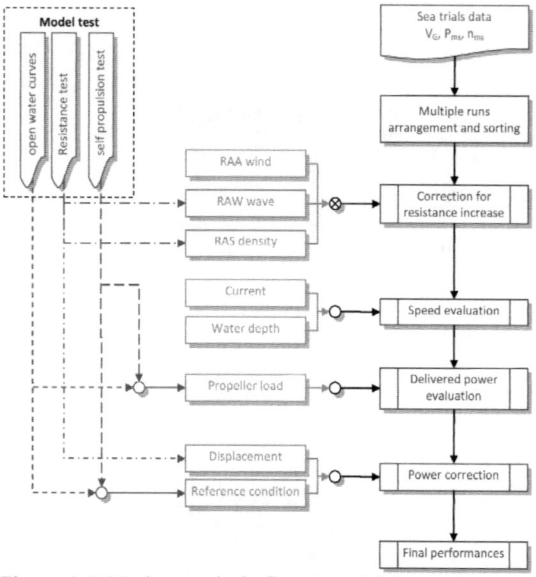

Figure 6: S&P data analysis flowchart (ISO 15016)

5. VIRTUAL WEATHER SERVICE

For the evaluation of the current weather conditions, the reference normative [1] prescribes the use of measurement devices like wave buoys and wave radar, the observations by multiple experienced mariners and the data provided by an experienced and independent meteorological office.
Despite anemometer is easy to be installed, it's usually difficult to have wave buoys available or wave radar onboard. For this reason, and in order to have uniform weather data set, CETENA decided to develop its own meteorological service, designed to be directly interfaced with the sea trials and monitoring systems in order to provide all the data needed for ship performance analysis. The system named VWS (Virtual Weather Station) is not directly interfaced with instrumentation/sensors measuring the weather conditions, but it is aimed to collect data from many weather/forecast services[1], to rebuild a single database with the more accurate spatial resolution and a time step equal to 1 hour. The data set considered are the results of simulations based on numerical models used for forecast and finally tuned with local direct measurements. For each sea/ocean the data set used for a complete description of the weather conditions needed for a naval approach are checked in advance, and made consistent with the database units and entities already considered. The VWS is currently configured in order to provide the following entities:

- Water depth
- Sea state
  - significant wave height
  - wave direction from
  - wave period
  - stokes drift components
- Current
  - Eastward component
  - Northward component
- Wind
  - Eastward component
  - Northward component

Furthermore, the current components can be provided at different draft, in order to identify the more significant effect on the ship motion. All the water entities describe the average value on one hour time step.
The decision of CETENA to develop a service for the weather condition prevision, is based on its experience on both sea trials and computational fluid dynamics. Some years ago, CETENA was asked to perform S&P sea trials on a ship in service, and then without the possibility to carry out the traditional multiple runs. On that occasion it has been decided to correct the ship speed using current data obtained by public weather services. The results obtained made it possible to reduce the spread of the ship speed values from 0.5 knots up to 0.3 knots, that is a value considered acceptable from all the other international partners involved in the project. So, stating the good results obtained, it has been decided to develop an own weather station, customized for a naval use to be directly interfaced with measurement/analysis systems. The VWS is developed in order to query and provide the required data directly to other systems requiring environmental conditions for their needs. Anyhow it is also possible to query the weather database for any other reasons using a custom user interface, shown in Fig. 7. This console gives the possibility of a direct investigation of weather conditions. Furthermore, in the case of the ship sailing conditions (speed and heading) are detailed for each required point, the apparent environment conditions are automatically computed showing the apparent wind and wave magnitudes and direction.

6. CERTIFICATION

The software SPEED has been checked and certified by two rules registers, ABS and RINA in order to guarantee the client both the robustness of the code development procedure and its compliance with the formulations prescribed by the normative.

6.1 ABS

ABS has been asked to check and certify the Software Development Life Cycle (SDLC) that guarantees the approach adopted by CETENA for the software development. As it can be seen in ABS rules [7], section 4-9-3:
"The Software Development Life Cycle (SDLC) cover the entire life of software development from concept to decommission of the computer-based system. The

[1] Copernicus – Marine environment monitoring service (marine.copernicus.eu)

software development process can be represented by a life cycle model consisting of several stages.[...]
The modified SDLC V-model is comprised of a sequence of stages that overlap and are iterative".

SPEED version 1.x.x.x is characterized by the ABS Certificate Design Assessment N. 17-GE1695772-PDA.

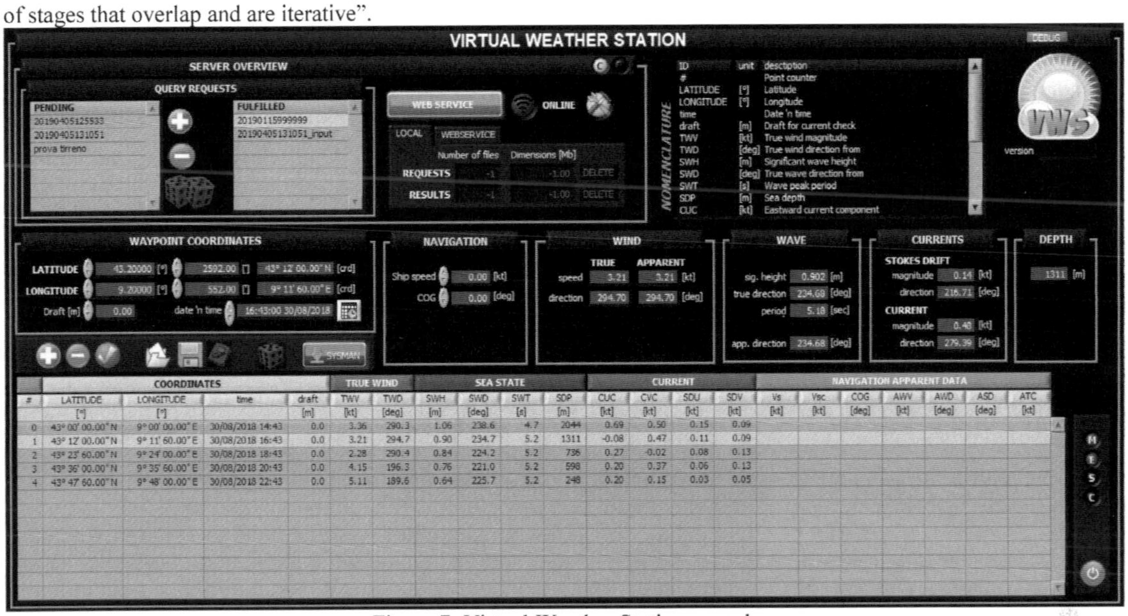

Figure 7: Virtual Weather Station console

6.2 RINA

The Italian Naval Register (RINA) has been asked to check and certify the accuracy of the results can be obtained by SPEED. After the selection of three test-cases, the results have been compared by RINA with the computation performed using their current analysis methods.
The test-cases used for the validation have been suggested by CETENA and accepted by RINA and are related to three different cruise ships, for which experimental data were available. The vessels considered are three ships with L_{BP} varying from 200 m up to 300 m length, and Δ in from 24000 t up to 78000 t.
The validation phase offered the opportunity to check and discuss with RINA technicians about the numerical approach suggested by the normative, the solution to be adopted for a correct implementation and its effect on the final results. Furthermore, together with RINA technicians the validation phase involved also discussions about the mathematical approach to be used, looking for the best compromise in terms of computation time request and accuracy. For example, different kinds of interpolation and integration approach have been investigated and tested.
The Italian Naval Register certified the results obtained through SPEED analysis are consistent with normative, through the RINA Statement N. MAC32418CS.
Recently RINA required a license in order to start in experiencing the software and use it for sea trial data analysis.

7. CONCLUSIONS

The possibility to identify the effective ship performances excluding the effects of wind, wave, load condition, deep water and salinity represents an important aspect of the sea trial execution, both from design and maintenances point of view. The official speed and power tests is a key point for the respect of the contractual agreed values and hence the delivery of the ship to the owner. The data analysis consistent with the last international normative and recommended procedures guarantees the owner about the computation of the effective ship performances and provide a good feedback to the shipyard from the design point of view. The development of the new CETENA software 'SPEED' makes it possible to finalize sea trials tests analysis and reporting directly onboard, using a very easy interface approach including the management of experimental data through a Microsoft Excel® spreadsheet file format, the possibility to query model test data and the check of the final power curve in contractual condition through custom consoles.
Just on board and in accordance with normative the owner, shipyard and sea trial company representatives have also the possibility to have at disposal, to see and to sign the outcome report assessing the ship official performance, including corrected measured data and the speed and power curve in contractual condition.

8. REFERENCES

1. 'Ship and marine technology – Guidelines for the assessment of speed and power performance by analysis of speed trial data', *ISO 15016*, 2015.

2. 'ITTC Recommended Procedures and Guidelines, Speed and Power Trials, Part-1 Preparation and Conduct', *ITTC 7.5-04-01-01.1*, 2014
3. 'ITTC Recommended Procedures and Guidelines, Speed and Power Trials, Part-2 Analysis of Speed/Power Trial Data', *ITTC 7.5-04-01-01.2*, 2014
4. 'ITTC Recommended Procedures and Guidelines, Preparation, conduct and analysis of Speed/Power trials', *ITTC 7.5-04.01-01.1*, 2017
5. 'Ship and marine technology – Measurement of changes in hull and propeller performance', *ISO 19030*, 2016
6. 'Rules for building and classing steel vessels – Part 4 Vessel systems and machinery', ABS, 2017

9. AUTHORS BIOGRAPHY

Paolo Becchi holds the current position of Senior Naval Architect at CETENA S.p.A., the research centre of Fincantieri Group. He currently works in the Ship Monitoring business unit since 2016, and he is involved in monitoring system and sea trial software development.
His previous experiences includes twelve years in CETENA Hydrodynamics dept., working on propeller design and analysis, Computational Fluid Dynamics through RANS computation and sea trials consultancy for full scale cavitation tests. He carried out military service at the Italian Navy Hydrodynamic Research Centre (Cavitation tunnel).

COMPUTATIONAL ANALYSIS TO CONTROL WELDING DEFORMATION IN THIN PLATE

Muhammad Taha Ali, EMSHIP, Belgium and Damen Shipyards Galati, Romania

SUMMARY

Globally in the shipbuilding industry, the computational method is a more productive method than the experimental and machine learning method. The Thermal Elastic Plastic (TEP) method for computational analysis is simulated in the FEMAP Software to analyse the welding deformation of thin plates. To validate the non – linear behaviour of the thin plate, the computational and experimental results are compared. The TEP – FEM method assesses the welding deformation in the thin plate and optimizes the welding procedure specification and sequence to increase productivity.

1. INTRODUCTION

1.1 BACKGROUND

In modern industry, new methods are evolving to increase the productivity of development stages in the industry. Modern manufacturing industry like automobile, train, aircraft and ship are accentuated to improve the fuel consumption, speed and carrying capability with capacity.

The shipbuilding industry is more focused to reduce the weight of the ship to decrease fuel consumption with an increase in efficiency. Decreasing the weight of the ship, thin plate structures are widely used in the ship structure especially in the passenger ships.

Welding technique is employed in the shipbuilding industry because this method of technique gives significant results and advantages. By welding techniques, different welding arrangement of the plates is analysed to join the profiles and plates. But during the welding process, the physical and chemical properties of the material of the plates and profiles are changed. That is why is called Multiphysics process. Due to Multiphysics process, the welding deformation occurs in the panel, section and block of the ship structure, which increased in the cost of the building, decrease in the productivity of work and delayed the production process.

Welding deformation problem is facing by the shipbuilding industry for seven decades and still, different mitigation techniques are followed to reduce the welding deformation. The most important mitigation technique that is applicable in the shipbuilding industry is the welding sequence method and the objective of this paper is focused on this mitigation technique.

1.2 OBJECTIVE

Control the welding deformation through the welding sequence of a transverse plating method.

1.3 PROBLEM

Shipbuilding industry still facing welding deformation problem after the welding process completed. This affects the cost of production and time duration of the project because more time is needed to straighten the plate. The industry is trying to control the deformation, but thin plate brings non-linear problems in the structure. This is due to different reasons, out of all three main reasons are described in this section.

One of the most important reasons for deformation in the thin plate is welding technique. For thick plates, welding robots are used for welding but for the thin plate, the manual technique is used and for this technique. For manual technique, different standards are used and skilled welders are needed for good quality welding. The second reason is the welding sequence, and it is changed with the thickness of the plate. The third reason which causes the welding deformation in the plate is the scantling of the panel. In this paper, all three problems which cause the deformation in the plate are considered to rectify it and develop the new methods to reduce it.

Thin plates normally welded near to the deck, superstructure and bulkhead of the hull structure in the passenger ships. Due to thin plate, low profile members are welded there. In past, the shipyard faces this problem due to scantling of the panel and then a lot of time is consumed to straighten the panel. This decreases productivity at the end of the project. As well as the outlook in the passenger ship is damaged. That is why in this paper bulkhead is considered for the analysis to rectify this problem.

Six types of welding distortion which causes the deformation in thin plates are given below [1]:

1. Longitudinal Shrinkage
2. Transverse Shrinkage
3. Angular Distortion
4. Bowing and dishing
5. Buckling
6. Twisting

It is possible that two or more than two distortions are effect at the same time.

For the experimental analysis, time and cost of the welding process increased and you to measure all the parameters during the welding process which consume a lot to time to complete the process. That is why computational analysis nowadays is a more useful tool to analysis the welding deformation.

1.4 CAUSE OF WELDING DEFORMATION

The main deformation causes in the thin plate are given below:

1. Heating
2. Human factor (like welder qualification etc.)
3. Movement of the plate
4. Weather effect
5. Fatigue
6. Degree of Restraint
7. The thermal distortion
8. Parent material
   - Inherent stresses included in previous metalworking processes.
     - Rolling
     - Forming
     - Bending
9. Design of the weldment
10. The accuracy of the manufacturer
11. Nature of the welding process
    - The symmetry of the joint
    - Number and sequence of welds required

1.5 METHODS TO CONTROL WELDING DEFORMATION

After welding deformation in the welding process, there are a lot of methods to control the deformation. Mainly, there are two main types of methods which help to control the welding deformation, which is given below:

1. Mechanical Solution
2. Thermal Solution

Other parameters of fixed problems are given below:

1. Welding Angles
2. Welding Sequence
3. Sequence of welding
4. Hammering
5. Mechanical Press
6. Thermal Straightening
7. Heat Treatment; relaxation
8. Shortening the welding time
9. Positioning and balancing the welds correctly round the axis.

In this research paper, deformation is controlled at the time of the welding process. That is why this method is a more useful and effective method to increase the productivity of the ship construction and computational analysis is a more productive method than other methods to analyse the welding deformation.

2. LITERATURE REVIEW

To control the deformation in a thin plate, a lot of research papers and thesis were performed experimentally and numerically. But still, this problem is faced by the researcher to give good agreements of the result to the shipyards.

Most of the researchers nowadays are working on thermal and mechanical response. This is because of the highly effective predicting tool Finite element simulation give the good result of thermo-mechanical behaviour. This method is performed in the computational analysis and it is a more powerful tool than other methods of analysis because it is more cost effective than the above methods of prediction.

That is why these days the FEM method is more commonly used in the welded structures to analyse. For Thermo-mechanical behaviour the analysis is performed in the FEM is Thermal Elastic-Plastic – Finite Element Method.

In researches, experimental and computational methods are used to analyse the deformation in the plate.

In 1954, Kumose et al. measure the angular distortion in the plate. Angular distortion is measured in the T – joint material. [2]

In 1997, Michaleris and DeBiccari measure the buckling and deformation in the plate in complex T – Joint fillet weld. [3]

In 2006, Tsai et al. integrated the numerical and experimental approach to investigate the model of welding induced buckling distortion process.

In 2008, Deng and Murakawa investigate the buckling distortion in thin plate structure by TEP – FEM and large deformation elastic FEM.

In their method following parameters are under consideration:

- Heat Input
- Thickness of plate
- Spacing Between stiffeners
- Welding Procedure Specification
- Welding Sequence

By these parameters buckling influence can investigate.

This Thermal Elastic-Plastic – Finite Element Method used in this research paper to investigate the welding deformation in the FEMAP software by Newton Raphson Method for non-linear analysis.

3. SCOPE OF CURRENT STUDY

Thermal mechanical uncoupled formulation: we can analyze the thermal and mechanical Behaviour of the structure. [4] In the welding process, the WPS i.e. welding procedure specification is a more important parameter to control the welding deformation in the plate. The values of the WPS are used in the computational analysis to validate with the experimental analysis in the production.

Welding procedure specifications have the parameters of mechanical and thermal analysis. Both analyses are performed separately that is why is called thermal mechanical uncoupled formulation.

Ten case studies are performed in this analysis. According to the uncoupled formation, there are two phases:

1. Thermal Analysis Phase and
2. Mechanical Analysis Phase

By using the mechanical analysis result, total deformation values are extracted and then it is validated with the experimental analysis deformation values.

4. MODELLING

The model assumes to be a temperature dependent. It means that thermal analysis is used to be a source of mechanical behaviour in the TEP-FEM analysis.

The following mechanical and thermal properties are the dependents of temperature are given below:

1. Yield Stress
2. Elastic Modulus
3. Thermal Expansion and
4. Poisson`s ratio.

For modelling of the plate need the values of temperature at the time of welding and the material of the plate and profile. The same heat source and the same material of the model are used for the analysis in thermal condition for a fillet weld. This analysis performed by Barsoum et al. [4]

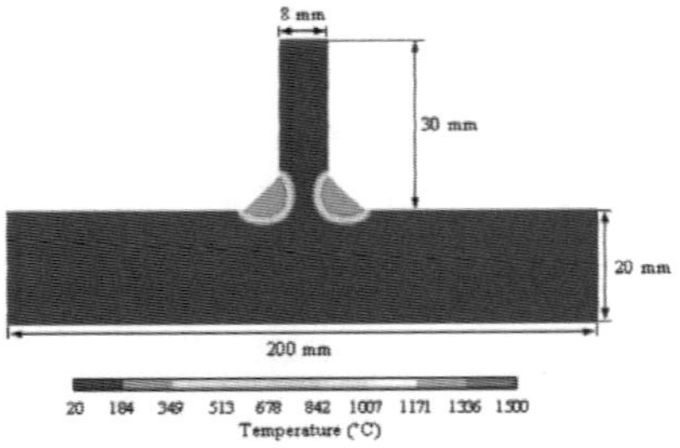

Figure 1: T-fillet welding performed to analysis thermal conditions in the thin plate. [5]

4.1 MATERIAL SELECTION

In the shipbuilding industry, two grade of steel is commonly used in the yacht building. First one is Grade "A" and the second one is Grade A36.

Grade "A" Steel is most commonly used steel in the shipbuilding industry for the thin plate in yacht and ferry boats. This steel is mild and hot-rolled. Welding properties of this steel are excellent.

4.2 COMPOSITION & PROPERTIES

In shipbuilding, it is necessary to align the plate properly at the beginning otherwise initial geometrical error cause the welding deformation.

The important parameter which is considered in the modelling is given below:

1. Density
2. Thermal coefficient
3. Thermal Conductivity
4. Thermal expansion coefficient
5. Specific Heat
6. Young modulus
7. Poisson ratio

From all the parameters few of them are temperature dependent parameters which are given below:

1. Density
2. Specific Heat
3. Thermal Conductivity

Other parameters are considered to be constant because the change in another parameter is very less.

Table 1: The chemical composition of Grade A steel. [6]

Element	Max. wt./ Content %
Fe	96.72
Mn	2.5
C	0.21
Si	0.5
P	0.035
S	0.035

Table 2: Mechanical properties of Grade A steel. [7]

Mechanical Properties	Metric
Modulus of Elasticity	200 GPa
Bulk Modulus	140 GPa
Poisson's Ratio	0.3
Shear Modulus	79.3 GPa
Yield Stress	235 MPa
Tensile Strength	400 – 502 Mpa

Table 3: Thermal properties of Grade A steel. [7]

Thermal Properties	Metric
Thermal Expansion	11 x 10^{-6} °C^{-1}
Thermal Conductivity	$0.084 \frac{W}{mm \times °C}$
Specific Heat	$480 \frac{J}{Kg \times °C}$
Initial Temperature	20 °C
Final Temperature	1500 °C

Cheapest and easiest welding of Grade "A" material is given below:

1. SMAW i.e. Shielded Metal Arc Welding
2. GMAW i.e. Gas Metal Arc Welding or MIG i.e. Metal Inert Gas Welding
3. Oxyacetylene Welding. [8]

4.3 MATERIAL PROPERTIES

Temperature is dependent on the two properties which are given below:

1. Thermo Physical Properties
2. Mechanical Properties

In the weld zone, heat source induced the high flux with a large temperature gradient. Due to the high temperature, the material property is changed radically.

The model in a constitutive manner is investigated in high temperature to predict the welding residual stress and welding deformation. [9][10]

Temperature depends properties are most important for thermal analysis as well as mechanical analysis. To give proper welding deformation analysis result. The process of the analysis is explained in the methodology.

5. METHODOLOGY

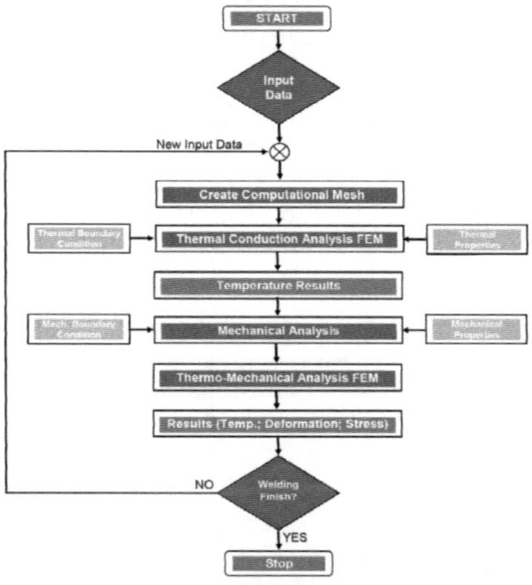

Figure 5: Research methodology flow diagram.

The computational research methodology flow diagram is schematically shown in the figure. In the flow loop methodology, there are three main modules which are:

1. Meshing
2. Thermal analysis and
3. Mechanical analysis.

Thermo-mechanical Analysis in FEM is in uncoupled mode between thermal and mechanical analysis. Both the analysis is done separately but the take the output of thermal analysis to use as an input of mechanical analysis.

In TEP – FEM analysis, FEMAP software used. Initially, modelling is done in FEMAP and the mapped meshing is performed. Refine meshing is taken into account for the thermal and mechanical analysis. This mesh helps to create nodes and elements. After meshing convergence of node has to be done to join the two plate properly. Meshing takes time if mesh size small then time will increase, and the computational speed is decreased.

In the second step in methodology flow is computational mesh to performed thermal analysis in FEM. Two nodes are in the corresponding relation and continue until the end of the plate to map the temperature from maximum temperature to lower temperature. The temperature at each node is very important that is why is save into the memory to use in a mechanical analysis as a thermal load. Because of this reason this analysis is known as thermos-mechanical analysis in FEM. Most of the computational time is consumed by the mechanical analysis in flow loop of methodology.

After this step in the result, we analyse the welding deformation and the residual stress in the structure. After the output of numerical results, must check the result validation. It is necessary to judge the result that it is correct or not. If yes, then the welding process is finished. Otherwise, the analysis process is returned to the starting point of the input data.

6. SIMULATION

In order to analyse the deformation in welded structure, a three-dimensional finite element model has been employed in FEMAP software. Different welding sequences are arranged to evaluate the best possible result.

FEMAP have has own modelling platform to design the 3D model.

6.1 FEMAP ANALYSIS & RESULT

For the thermal and mechanical analysis, FEMAP software is used and FEMAP NASTRAN is used as a Solver. For Simple and complex calculation this software is accurate and friendly user interface software.

Model of the structure is built in FEMAP. After the simulation FEMAP have an option to formed, contours and plots. [11]

This research paper performed ten case studies simulation to compare the best result of the transverse plating method to normal bulkhead method.

For the non-linear analysis in FEMAP software used the Newton Raphson Method. The maximum value of the stress and total deformation which are simulated and it`s varies from one case study to another case study due to the transverse plating method and fixed boundary conditions.

Due to this parameter in computational analysis more refined and accurate mesh should be performed. Otherwise, the wrong result can be generated in the FEM analysis.

Thermal loads are used in the form of temperature or heat flux in FEM analysis. The result shows that the maximum thermal load near the welding toe of the fillet welds.

7. RESULT COMPARISON OF SIMULATION

Due to the welding process and it`s complexities require to validate the analysis by welding deformation result. Welding complexities in the process is due to the temperature distribution in the model and the material properties which are depended on temperature. That is why in different case studies in this paper are performed to focus on the temperature distribution in the model to be uniform because of thermal elastic-plastic finite element method. In this method, uniform distribution effects of the thermal load affect the residual stress especially deformation.

That is why ten case studies are performed to analyse the best possible result of welding deformation. By comparing all case studies, one case study gives the best possible result of welding deformation and residual stress. In this case study, the corner and the centre of the base plate are fixed.

These case studies are performed in the bulkhead panel of the ship. In these case studies, both normal bulkhead panel and modified panels are analysed in FEMAP software by using the TEP-FEM method.

The total deformation in normal bulkhead panel is shown in the below figures:

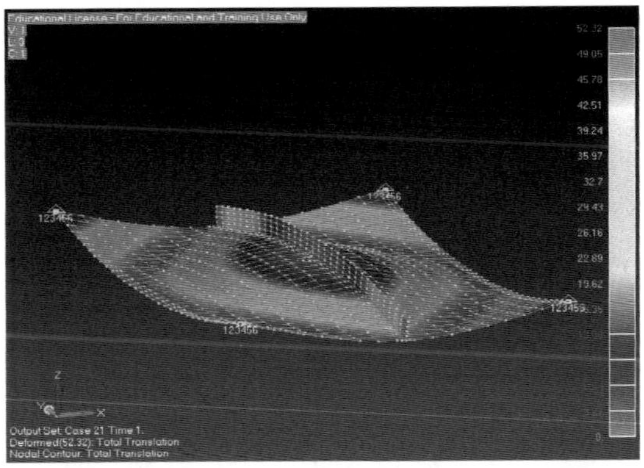

Figure 6: Total deformation result in normal bulkhead panel

The total deformation in modified bulkhead panel is shown in the below figures:

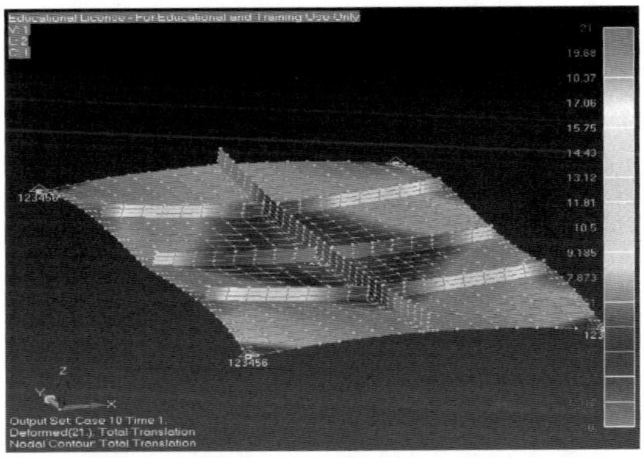

Figure 7: Total deformation result in modified bulkhead panel

From figure 6 and figure 7 it is clearly shown that welding deformation in figure 6 is much higher than figure 7. This is because of the transverse Plating method with continuous welding which helps the temperature distribution is more uniform. Due to uniform temperature distribution, thermal load in the model is also uniform which helps in the structural analysis to find out the welding deformation. Due to the uniform thermal load in a model reduce the welding deformation as compared to case study # 1 in which model is fixed from the centre and the corners of the model.

After computational analysis, a thin plate of the bulkhead shows the good agreement of results to reduce the welding deformation of the structure by TEP – FEM. Which means that transverse plating arrangement of the flat bar can control the welding deformation in the thin plate. This analysis shows that the solution is highly efficient and accurate to control the welding deformation in the thin plate.

8. VALIDATION

To validate the thermal elastic-plastic finite element method with the experimental method in the thin plate by the welding deformation displacement results.

Computational Result

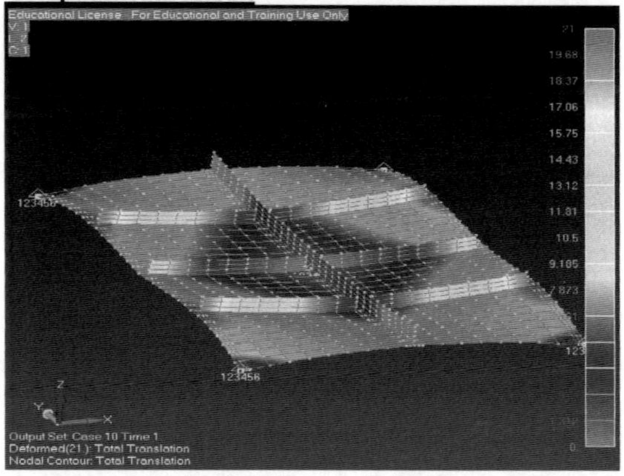

Figure 8: The computational result of total deformation.

In the computational result, the maximum total welding deformation is 21mm in this case study. When the model is fixed with the centre and corners.

Experimental Result

Figure 9: Experimental Case study when the model is fixed at the centre and corners of the plate.

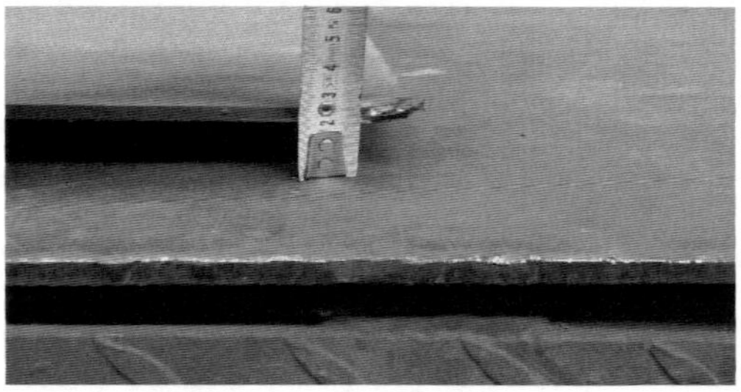

Figure 10: The Experimental result of total deformation.

In the experimental result, the maximum total welding deformation is 20.5 mm in this case study when the model is fixed with the centre and corners.

From figure 8 and figure 10, it is clearly shown that welding deformation error between both the results is **2.4%**. It means that both the results are validated by each other. This is because of transverse fairing method with proper welding sequence to distribute the temperature in the panel, welding procedure specification to make a good size of welding bead and continuous welding which helps the temperature distribution is more uniformly.

In the experimental analysis, different sequences are performed to minimize the welding deformation in the plate and clamping method also takes into account.

By consideration, the welding deformation is reduced in the case study when the model is fixed with the centre and corners in computational analysis and experimental analysis.

Welding deformation are required for the welding process complexity, this includes the following parameters:

- Localized heating
- Clamping
- Temperature dependence of material properties.

9. CONCLUSION

This paper developed the method to control the welding deformation in the thin plate is more accurate and productive through computational analysis. This method becomes more relevant to the other methods of fairing after the welding process because this method is performed at the time of welding process which increases the productivity of the production process. To satisfy why this method is more accurate and useful, consider one fairing method which is mostly used in the yacht building that is the steel putty method. This method increases the cost of production and also increases production time. And without transverse plating method 60% of welding deformation increase, means more time is required for straightening and more steel putty required for the good outlook of the ship.

That is why the transverse plating method at the time of the welding process gives a good agreement of productivity. This method gives the good agreement of result in Computational analysis and experimental analysis. And both the analysis are validated each other through **2.4%** error which is neglected error. So, Computational analysis is good for further studies in a complex and large structure.

The productivity in the bulkhead panel is increased up to **12 – 14%** by this method with respect to the total time of production and labour cost.

This result helps to analysis the large and complex structure of the ship.

10. RECOMMENDATIONS & FUTURE WORK

To control the welding deformation during the welding process, the following methods in the future can be used in shipyards:

- Clamping method;
- Welding procedure sequence,
- Different welding sequence.

Shipyard must focus on the experimental and computational analysis to validate the new methods such validated methods may increase in the future the productivity and accuracy of the project.

After the welding process, the control welding deformation inductive fairing method can also be implemented in shipyards. These methods and techniques will be useful for the shipyard to control the deformation.

11. ACKNOWLEDGEMENTS

This study has been accomplished in the framework of EMSHIP – European Masters Course in Advanced Ship and Offshore Design (www.EMSHIP.EU), coordinated by University of Liege, Belgium. I acknowledged the support by the Research Centre of Naval Architecture Faculty from University of Galati and Damen Shipyards Galati.

12. REFERENCES

1. MASUBUCHI K, Analysis of welded structures. *Pergamon Press, USA.*

2. KUMOSE T., YOSHIDA T., ABBE T. and O. H, Predicting of angular distortion caused by one-pass fillet welding. *Welding Journal (vol. 1) 945 – 56*, 1954.

3. DEBICCARI M., Prediction of welding distortion. *Welding Journal (vol. 1) 172 – 81*, 1997.

4. MEHDI GHANADI, Managing distortion in welded structures using FEM. Master Thesis, 2013.

5. BARSOUM I. B. Z., Residual stress effects on fatigue life of welded structures using FEM, *Engineering failure analysis (no. 16) 449 – 467*, 2009.

6. www.shanghaimetal.com/shipbuilding_plate-129.htm

7. IACS Rule # 47.

8. PAZOOKI A.M.A, HERMANS M.J.M., RICHARDSON I.M., Control of welding distortion GMAW of AH36 plates by stress engineering, *Int J Adv Manuf Technol 88:1431457,* 2017.

9. HUI HUANG, Large Scale Thermal Elastic-Plastic Analysis Based on Dynamic Mesh Refining Method and Iterative Substructure Method, 2016.

10. QIAO D, ZHANG W, FENG Z., High-temperature constitutive behaviour of austenitic stainless steel for weld residual stress modelling. *ASME Pressure Vessels and Piping Conference. American Society of Mechanical Engineers, 1229-1236*, 2012.

11. HASSAN MUNEEL SYED, Modelling of residual stresses and distortion due to welding in fillet welds, *Master Thesis*, 2013.

13. AUTHORS BIOGRAPHY

Muhammad Taha Ali, EMSHIP, holds the current position of Research Intern at Damen Shipyards Galati. He is responsible for developing new methods to control deformation in the shipbuilding process.

AUTHOR INDEX

Ali, M. 55
Becchi, P. 48
Boulougouris, E. 1
De Meij, R. 8
Egorov, A. 20
Egorov, G. 20
Grin, R. 8
Hudson, D. 36
Ilnitskiy, I. 20
Kamath, Y. 16
Nunes, D. 36
Rapuc, S. 8
Sivaprasad, K. 16
Stefanidis, F. 1
Vassalos, D. 1
Wye, A. 29

The Royal Institution of Naval Architects
8-9 Northumberland Street
London, WC2N 5DA
United Kingdom

ISBN 978-1-5108-9540-9

CURRAN ASSOCIATES INC.
proceedings.com

Research and Education 2018

Transportation Research Record: Journal of the Transportation Research Board

Volume 2672, Issue 13

SAGE Publishing